D0437867

Benchmarking
for
Competitive Advantage

Robert J. Boxwell, Jr.

McGraw-Hill, Inc.

New York   San Francisco   Washington, D.C.   Auckland   Bogotá
Caracas   Lisbon   London   Madrid   Mexico City   Milan
Montreal   New Delhi   San Juan   Singapore
Sydney   Tokyo   Toronto

LIBRARY-MEDIA CENTER
SOUTH PUGET SOUND COMM. COLLEGE
2011 MOTTMAN RD. S.W.
OLYMPIA, WA 98512-6292

WITHDRAWN

Library of Congress Cataloging-in-Publication Data

Boxwell, Robert J.
    Benchmarking for competitive advantage / Robert J. Boxwell, Jr.
      p.   cm.
    Includes index.
    ISBN 0-07-006899-2 (alk. paper)
    1. Benchmarking (Management).   2. Organizational effectiveness.
    3. Performance.   4. Competition.   I. Title.
    HD62. 15.B69   1994
    658.5'62—dc20   94-3940
                                                          CIP

Copyright © 1994 by McGraw-Hill, Inc. All rights reserved. Printed in the United States of America. Except as permitted under the United States Copyright Act of 1976, no part of this publication may be reproduced or distributed in any form or by any means, or stored in a data base or retrieval system, without the prior written permission of the publisher.

2 3 4 5 6 7 8 9 0   DOC/DOC   9 0 9 8 7 6 5

ISBN 0-07-006899-2

The sponsoring editor for this book was Harold B. Crawford, the editing supervisor was Stephen M. Smith, and the production supervisor was Pamela A. Pelton. It was set in Palatino by McGraw-Hill's Professional Book Group composition unit.

Printed and bound by R. R. Donnelley & Sons Company.

 This book is printed on recycled, acid-free paper containing a minimum of 50% recycled de-inked fiber.

Information contained in this work has been obtained by McGraw-Hill, Inc., from sources believed to be reliable. However, neither McGraw-Hill nor its authors guarantee the accuracy or completeness of any information published herein and neither McGraw-Hill nor its authors shall be responsible for any errors, omissions, or damages arising out of use of this information. This work is published with the understanding that McGraw-Hill and its authors are supplying information but are not attempting to render engineering or other professional services. If such services are required, the assistance of an appropriate professional should be sought.

In memory of my Dad, who only knew this book as piles of paper cluttering up the dining room table during my vacations home. As a person and as a father, Dad was the benchmark.

047199

Contents

Preface

*Give us the tools, and we will finish the
job.* WINSTON S. CHURCHILL

A telephone company manager, interested in the efficiency of
her sales agency support group, contacts her counterparts at the
other six regional Bell operating companies to find out their
ratios of sales agents to support group personnel and other
measures she considers meaningful.

Is this benchmarking?

The head of North American Manufacturing at General
Motors wants to compare his labor force's efficiency with those
of Ford, Chrysler, and Honda. He does some research and finds
the exact measures he is seeking in a recent issue of *Time* magazine.

Is *this* benchmarking?

A group of senior executives from a large credit card company visits USAA for a day to discuss a wide range of issues related to quality. This benchmarking trip was arranged by the vice
president of quality at the company, who called his counterpart
at USAA and set up the visit in a couple of brief phone conversations. Twelve senior executives went to San Antonio and

spent the day as a group with an approximately equal number of USAA executives, who gave their "quality briefing."

How about this? It was a sunny day in San Antonio. Is this benchmarking?

Three managers from the same credit card company decide to "benchmark" their training operations. The whole thing. Although they already have a "superior training capability" and "don't have a lot of metrics" by which to compare themselves to other companies, they manage to convince six other companies renowned for their training functions to participate in an ad hoc consortium to study training. The three managers develop a detailed questionnaire and distribute it to each of the participants, collect all the responses, and compile the responses in a Participant's Report, which they distribute to the other participants. They also follow up with further requests for data and conference calls to learn more about areas of particular interest to them based on the participants' responses to the questionnaire. The managers then encourage the other six participants to do likewise.

And is this benchmarking?

What about this next one?

A senior manager at Ford is told to design a new mid-size car line to win back some of Ford's lost market share. He studies over 400 features of Japanese, *West* German (no Trabants), Swedish, and U.S. cars, and his team ends up with the Taurus and Sable models.

What do you think? Benchmarking?

A ready-mix concrete company that is losing a large number of competitive bids to its two primary competitors retains an outside consulting firm to "benchmark its competitors' costs." Although managers from the company provide a great deal of input to the study, they do not have any contact with their competitors during the study—the outside consulting firm "interviews" the competitors.

Is this benchmarking? Is it ethical?

A vice president of a large chain of movie theaters takes her husband and children to Disneyland for the weekend to "benchmark how Disneyland's facilities management handles crowd control."

You would probably like to be part of this "site visit," but would you be benchmarking?

The manager responsible for category 6 of his company's Baldrige Award application hires an outside benchmarking firm to gather data on various quality measures in the computer and computer-related industries. These "metrics" will be shown vis-à-vis his company's results in its Baldrige application.

Last chance. Is this benchmarking?

If you answered yes to all the above, you would be right, sort of. Each of the above is an actual example of what people call benchmarking. And although some of these are pretty weak examples of benchmarking, one cannot blame these managers for calling what they were doing benchmarking, because it has become an intellectually fashionable term over the past few years. To say you are using benchmarking to solve some problem or enhance some process means you are using what is one of the most powerful and sensible management improvement processes to come along in quite some time. But are you doing it correctly? Is your benchmarking rigorous and valuable? Or is it industrial tourism? If your benchmarking is intellectually rigorous, that's great. You will likely reap great rewards from it, which will include making your organization more competitive, educating yourself, and being considered smart by your colleagues. If your benchmarking is industrial tourism, that, too, may be okay, as long as you have made the conscious decision to perform "benchmarking" of the industrial-tourism variety. There's probably more value to phoning a couple of companies to have a brief conversation about how they do something than to sitting in your office and not picking up the phone. But you need to know that such an exercise barely, if at all, qualifies as real benchmarking. It's when you think you are performing benchmarking correctly (i.e., rigorously) and you are really only being an industrial tourist that you may find yourself in deep trouble.

This book is written for people who want to learn what real benchmarking is and hit the ground running with their newly acquired knowledge. This book is easy to follow and contains lots of examples of how managers in a wide variety of industries have used benchmarking. And it is laid out in a logical sequence, which largely parallels the sequence of events you would follow when planning and executing your study and using your findings to improve your organization. I hope you enjoy reading it. I have tried to keep it enjoyable—some sarcasm, irreverence, and humor are sprinkled throughout. I also hope you will find it helps you in your benchmarking endeavors.

Robert J. Boxwell, Jr.

Acknowledgments

Many, many clients, friends, and family members contributed to the completion of this book. I hope my memory serves me well in acknowledging them all here. I am extremely grateful to all of them for their contributions.

First, tremendous editorial and content help was given to me by Bert Nelson, Gene Ceccotti, and George Pavlov. They combined to take a rudimentary manuscript and turn it into a book. If anything in this book sounds stilted or unclear, it was doubtless written by me alone *after* they provided their invaluable input.

Since I began benchmarking several years ago, I have found my way onto the mailing lists of many friends who feel it is their obligation to send me each and every benchmarking or benchmarking-related article that crosses their desks. I have received scores in the last year alone from Suzanne Aquino, J. Edgar Klinge, Lelyana Kurniawan, Ron Laher, Sheila Lee, Robert Lamb, Lori Meisinger Lupton, Carter Mario, Dean McCaskill, Scott Ragaglia, Joe Roberts, Dave Snyder, Matt Wemple, and others.

Other intellectual input has come from many sources. First, Andrall Pearson of Harvard Business School took time out from a vacation two years ago to educate me on some of the history behind the concepts and practice of strategic planning. Several classmates of mine, Dave Buzby, Teresa Clarke, Leigh-Ellen Louie, Loy Sheflott, and Susheila Vasan, provided other valuable intellectual food for which I am sincerely grateful. Jeff Rumburg provided me with encouragement during my early benchmarking days. Some members of the Churchill & Company family, Claire Chen, John Kinnaman, Dora Lee, and Betty Silva, also contributed significantly to research and examples that are sprinkled throughout this book. And thanks to my family, Nan, Dad, Mom, Audrey, Mimi, Malcom, Felipe, and Steve, who all provided moral support throughout the project. Malcom also endured some of my poor handwriting and contributed to the typing.

Which brings me to Cecilia Tabaranza-Dioso, who really produced this book. Cez spent hundreds of hours deciphering my scratch, making changes, correcting my misspellings, and doing what would have taken me years to complete. To her I give special thanks. I would still be on Chap. 3 if she weren't around.

1

A Brief History of Strategic Planning and the Evolution of Benchmarking

Advice to persons about to write history—don't.   LORD ACTON

Introduction

Strategic planning (also known as long-range planning) began to gain widespread momentum and popularity in the 1960s. Many of the more popular strategic planning tools provided insightful frameworks through which managers could think about issues and challenges facing them at the corporate strategy level. The insight gained from using these tools was often powerful and could result in development of strategies and subsequent implementation that would affect many companies and the lives of their employees. A decision to divest a subsidiary in an unattractive industry, e.g., could affect the lives of thousands of people.

But the fact that much of the insight from these frameworks was at a "big-picture" level helped create the demand for a process like benchmarking. Needed was a process that could be used by managers *throughout* an organization to improve their areas of responsibility and hence the competitiveness of the organization. Line managers could become involved—in fact, should be involved—and could benefit from benchmarking by learning how to make improvements and *how* to execute at a tactical level. As the name implies, most strategic planning tools addressed only strategy and considered very little about execution. The tools did not tell managers *how* to do anything. This, of course, does not imply that they were not powerful managerial tools. Simply, they were not always tools that could be put to good use by managers throughout the entire organization.

This chapter discusses a brief history of strategic planning and the evolution of benchmarking. It may not be of interest to those who wish to jump right into the mechanics of the benchmarking process, but it may be valuable for those who wish to put benchmarking in perspective to ensure the ultimate success of an organizationwide benchmarking process. Benchmarking itself is not necessarily a strategic planning tool, but it fits into the strategic planning process at the juncture of planning and execution. As you consider some of the big-picture frameworks on the following pages, you will understand why it was only a matter of time before benchmarking became popular in the United States.

A Brief (*Very* Brief) History of Strategic Planning

To understand how benchmarking became so popular and gained such wide acceptance by U.S. managers, you should be familiar with some of the history of strategic planning as a discipline and the tools developed to enable managers to analyze strategic issues. The strategic planning frameworks covered on the next several pages were, in most cases, high-level analytical tools, useful for making macro-level decisions about the big strategic issues most often facing large corporations.

Strategic planning gained momentum during the 1960s and reached its peak during the 1970s. In the 1960s, many companies began what was called *long-range planning*. Thinking ahead seemed like a good thing to do. Much of the process was driven by mathematical models, many of which were descendants of models originally developed by the Office of Strategic Services during and after World War II. These models were designed to determine what outputs should be obtained from a given set of inputs. *The Economist* (March 20, 1993, p. 76) describes the phenomenon:

> By the 1960s corporate strategy had come to mean a complex and meticulously wrought plan based on detailed forecasts of economies and specific markets. That view was endorsed by two celebrated books: Alfred Sloan's "My Years With General Motors," a memoir by the man who made the car maker the world's biggest industrial enterprise; and Alfred Chandler's "Strategy and Structure," a history of big, successful American firms in which the Harvard professor argued that their strategies had produced their multi-divisional form.
>
> This approach to strategy fell into disrepute for several reasons. Many people blame it for the over-zealous diversification of the following decade and the creation of poorly performing conglomerates. In the 1970s the success of Japanese firms, which seemed to eschew detailed planning, cast further doubt on its usefulness. The two sudden oil-price rises of the 1970s also meant that many firms had to tear up their plans and start again. Most revealing of all, many companies found that the reams of statistics and targets, once assembled, sat gathering dust. Occupied with running their operations, few managers at any level of the firm ever bothered to refer again to its handsomely bound corporate strategy.

At about the same time, a few of the management consultancy firms in the avant garde developed some ground rules and strategic planning tools to enable their large corporate clients to better understand corporate strategy issues facing their diversified businesses. Two of the more popular and easy-to-understand analytical tools were the Boston Consulting Group's

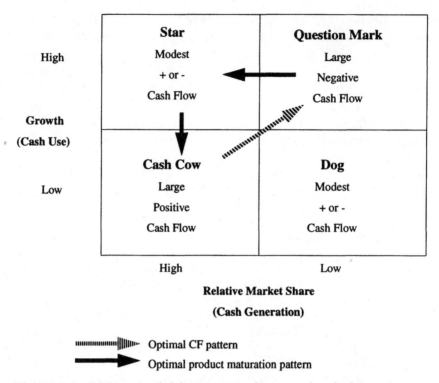

Figure 1-1. BCG's growth/share matrix. (Reprinted with the permission of The Free Press, a Division of Macmillan, Inc. from *Diversification Through Acquisition: Strategies for Creating Economic Value*, by Malcolm S. Salter and Wolf A. Weinhold. Copyright © 1979 by Malcolm S. Salter and Wolf A. Weinhold.)

(BCG) *growth/share matrix* (Fig. 1-1) and the McKinsey *nine-box matrix* (Fig. 1-2). They are remarkably similar in terms of analysis provided but sufficiently different so as not to be confused by corporate managers entertaining proposals from competing consultancies for big think-piece engagements.

BCG's growth/share matrix and McKinsey's nine-box matrix were tools that enabled managers to analyze a portfolio of companies under common ownership to determine the optimum flow of resources (usually cash and senior management attention) among them. When used as competitor analysis tools, they provided an analyst with the ability to determine where in a parent's portfolio a particular business unit lay, from which

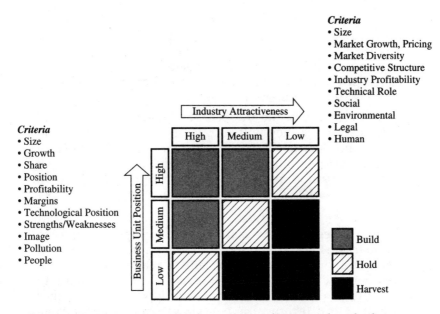

Figure 1-2. McKinsey's nine-box matrix. (Reprinted with the permission of The Free Press, a Division of Macmillan, Inc. from *Competitive Strategy: Techniques for Analyzing Industries and Competitors* by Michael E. Porter. Copyright © 1980 by The Free Press.)

inferences could be drawn about the parent's likely response to competitive moves. For example, if a competing business were determined to be a "dog" under the BCG portfolio, it might be reasonable to believe that a rational parent would be harvesting that business and, consequently, might not respond vigorously to battles for market share. The same example applies to businesses determined to be in a harvest mode by using the McKinsey nine-box matrix.

The growth/share matrix is very simple and easy to understand.

- *Stars* in the upper left quadrant are the investment opportunities. With high cash use and cash generation (due to a favorable cost position in its industry), they are relatively self-sufficient.

- *Cash Cows* in the lower-left quadrant generate high cash flows yet use little in their low-growth market. These are net providers of cash, which is typically funneled to Question Marks, and R&D projects.

- *Dogs* in the lower-right quadrant are cash traps, in that additional cash investments cannot be recovered. Increasing market share in a stable market is futile because no one can afford to increase capacity any more than they can afford to run at much less than full capacity. Dogs are candidates for liquidation.
- *Question Marks* in the upper-right quadrant are the real risks. Left alone they will become dogs as market growth slows, and their profit margins contract relative to those of the industry's dominant competitors. To make stars of them requires a great deal of cash for investment in increased market share and, therefore, accumulated experience.

It should be emphasized that the relationships among cash flow, market share, and market growth are likely to be most evident in industries where (1) there is a substantial experience phenomenon and (2) lower costs can be translated into competitive advantage either through lower prices, higher marketing and technical expenses, or some other means. Where lower costs are not critical to success, such as in highly differentiated markets or where experience-curve effects are limited in time because of a high rate of technological change or when product obsolescence occurs, these relationships may be tenuous.[1]

The BCG growth/share matrix relies heavily upon the existence of an experience curve in a cash cow's industry. If an experience curve exists, the market share leader should, over time, develop a cost advantage vis à vis its competitors. Because of its inherent reliance on qualitative factors in its assessment of business unit position and industry attractiveness, the McKinsey nine-box matrix is somewhat more subjective than the growth/share matrix. Essentially, however, it provides a similar type of strategic insight.

Both these frameworks were powerful tools to develop planned actions at a boardroom level. Much of their power as communication tools lay in their simplicity and ease of applica-

---

[1]Reprinted with the permission of The Free Press, a Division of Macmillan, Inc. from *Diversification Through Acquisition: Strategies for Creating Economic Value* by Malcolm S. Salter and Wolf A. Weinhold. Copyright © 1979 by Malcolm S. Salter and Wolf A. Weinhold.

bility. A two-by-two matrix is worth a thousand words. The problem is, those thousand words do not say much about *how* to do anything. Firms that wished to address this next level of detail, i.e., how to improve, in the early stages of the long-range planning movement often got caught up in corporatewide improvement efforts that tended to spread their resources too thinly. In a sense, they were trying to do *everything* better, and usually they achieved the results one might expect when such an approach to improvement is taken.

Both these frameworks could be misleading or not applicable at times, too. The BCG growth/share matrix was partly founded on the belief that sustainable competitive advantage could be gained through moving down the experience curve, which is not always the case. And the McKinsey nine-box matrix was useful in addressing strategy at a business unit level but not at a holding company level.

Another popular strategic planning framework was the *product life cycle,* which Michael Porter, a Harvard Business School professor, calls "the grandfather of concepts for predicting the probable course of industry evolution."[2] Under this concept, an industry or a product passes through four phases: introduction, growth, maturity, and decline. The concept, still used, was not universal but applied to a large enough number of industries and products that it was applicable quite frequently. A graphical depiction of the life cycle is shown in Fig. 1-3.

Toward the end of the 1970s, Porter began work on the first of his books on industry analysis and how firms compete with one another. *Competitive Strategy* was called "one of the ten best management books ever written" by Tom Peters, of *In Search of Excellence* fame. *Competitive Strategy* introduced, in succinct, understandable form, what had previously been diverse and scattered concepts for analyzing industries and the competitors within them. Porter's five forces and three generic strategies shed much light on industry analysis and methods of competing, all distilled to an easy-to-understand format.

---

[2]Reprinted with the permission of The Free Press, a Division of Macmillan, Inc. from *Competitive Strategy: Techniques for Analyzing Industries and Competitors* by Michael E. Porter. Copyright © 1980 by The Free Press.

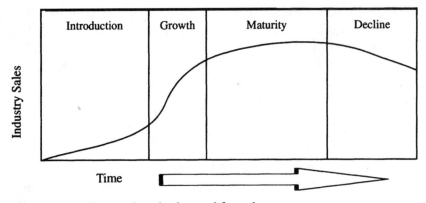

Figure 1-3. The product/industry life cycle.

According to *The Economist* (March 20, 1993, p. 76), Porter "argued that a firm's profitability was determined by the characteristics of its industry and the firm's position within it, so these should also determine its strategy. Applying the analytical techniques common to industrial economics, Mr. Porter said that a firm's primary task was to find niches it could defend from competitors, either becoming the low-cost producer, differentiating its products in a way which would allow it to command a higher profit margin, or erecting barriers to the entry of new rivals."

Porter's *five forces,* shown in Fig. 1-4, are the forces that affect the profitability of practically any industry. A sophisticated understanding of the five forces and their effect on an industry provides keen insight to the overall profitability and future potential for profitability of that industry. In any analysis of a competitive situation, analysis of the overall industry is the logical first step.

Within an industry, firms could compete in different ways. Those with the most successful strategies should be fortunate enough to earn above-average returns compared to other firms in the industry. Porter summed it all up very simply by saying that firms in an industry could choose one of three generic strategies by which to compete (see Fig. 1-5):

■ Overall industrywide cost leadership

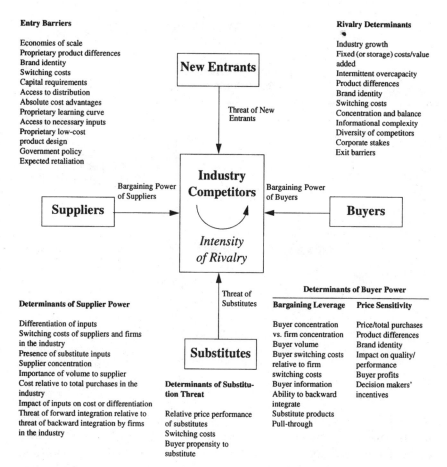

Entry Barriers

Economies of scale
Proprietary product differences
Brand identity
Switching costs
Capital requirements
Access to distribution
Absolute cost advantages
Proprietary learning curve
Access to necessary inputs
Proprietary low-cost
product design
Government policy
Expected retaliation

Rivalry Determinants

Industry growth
Fixed (or storage) costs/value added
Intermittent overcapacity
Product differences
Brand identity
Switching costs
Concentration and balance
Informational complexity
Diversity of competitors
Corporate stakes
Exit barriers

Determinants of Supplier Power

Differentiation of inputs
Switching costs of suppliers and firms in the industry
Presence of substitute inputs
Supplier concentration
Importance of volume to supplier
Cost relative to total purchases in the industry
Impact of inputs on cost or differentiation
Threat of forward integration relative to threat of backward integration by firms in the industry

Determinants of Substitution Threat

Relative price performance of substitutes
Switching costs
Buyer propensity to substitute

Determinants of Buyer Power

Bargaining Leverage	Price Sensitivity
Buyer concentration vs. firm concentration	Price/total purchases
Buyer volume	Product differences
Buyer switching costs relative to firm switching costs	Brand identity
Buyer information	Impact on quality/performance
Ability to backward integrate	Buyer profits
Substitute products	Decision makers' incentives
Pull-through	

Figure 1-4. The five forces. (Reprinted with the permission of The Free Press, a Division of Macmillan, Inc. from *Competitive Advantage: Creating and Sustaining Superior Performance* by Michael E. Porter. Copyright © 1985 by Michael E. Porter.)

- Industrywide differentiation
- Focus, using cost leadership or differentiation in a particular market segment only

In essence, a firm could be the overall low-cost producer in an industry and, by virtue of that position, earn above-average returns in the industry. Or, if not the low-cost producer, a firm could differentiate its products and/or services somehow from

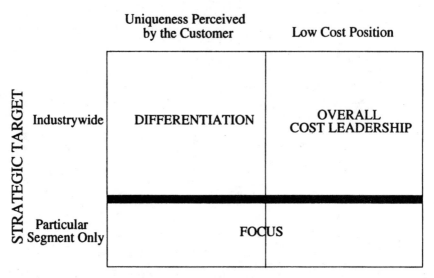

Figure 1-5. Three generic strategies. (Reprinted with the permission of The Free Press, a Division of Macmillan, Inc. from *Competitive Strategy: Techniques for Analyzing Industries and Competitors*, by Michael E. Porter. Copyright © 1980 by The Free Press.)

those of other competitors in order to charge a higher price for them. Above-average profits could be achieved by a differentiating firm, too, as long as the cost of differentiation was less than the premium that the firm was able to command in price for its differentiated products or services.

Once a manager understood industry profitability and the potential strategies that could be employed, he or she could then dissect an industry and develop a thorough competitive analysis to be used in strategy formulation. The growth/share matrix or the nine-box matrix might be part of this analysis. *Competitive Strategy* provides additional frameworks for analyzing competitors, including the beliefs-goals-capabilities-assumptions framework and strategic maps.

By using the beliefs-goals-capabilities-assumptions framework (Fig. 1-6), an individual competitor can be analyzed at a macro level, and educated guesses can be made about how

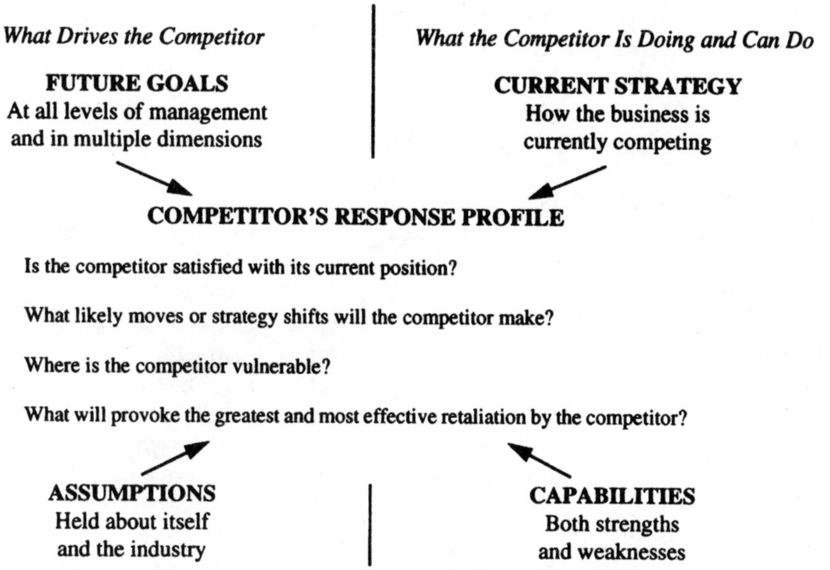

What Drives the Competitor | *What the Competitor Is Doing and Can Do*

FUTURE GOALS
At all levels of management
and in multiple dimensions

CURRENT STRATEGY
How the business is
currently competing

COMPETITOR'S RESPONSE PROFILE

Is the competitor satisfied with its current position?

What likely moves or strategy shifts will the competitor make?

Where is the competitor vulnerable?

What will provoke the greatest and most effective retaliation by the competitor?

ASSUMPTIONS
Held about itself
and the industry

CAPABILITIES
Both strengths
and weaknesses

Figure 1-6. Components of analysis of a competitor. (Reprinted with the permission of The Free Press, a Division of Macmillan, Inc. from *Competitive Strategy: Techniques for Analyzing Industries and Competitors* by Michael E. Porter. Copyright © 1980 by The Free Press.)

one's own firm can best compete with that competitor. Porter makes the point that "most companies develop at least an intuitive sense for their competitors' current strategies and their strengths and weaknesses" (shown on the right side of Fig. 1-6), but that "much less attention is usually directed at the left side, or understanding what is really driving the behavior of a competitor—its future goals and the assumptions it holds about its own situation and the nature of its industry."[3]

A McKinsey view of competitive analysis is as follows:

Conventional competitive analysis is usually conducted as three largely independent streams of analysis that are later integrated and cross-checked to establish a unified picture of what competitors actually accomplish. Reverse product

[3]Ibid., p. 48.

engineering is an important part of most analyses. It highlights effective product designs, their costs, and the process technology necessary to produce them. A thorough financial analysis of publicly available information provides further evidence of competitive cost positions by focusing on total business system economics.

Financial analysis can provide an indication of capacity utilization, illustrate the strengths and weaknesses in alternative paths to market, and highlight differences in the way competitors have decided to concentrate resources. Finally, these exercises are supplemented with field work—interviews with suppliers, distributors, and customers—to fill in gaps in knowledge and to provide tangible illustrations of how different approaches are received in the marketplace.

Competitive analysis is especially useful when top management is faced with a major strategic choice—acquisition or divestiture, make/buy, entry/exit, or business restructuring. Detailed and rigorous competitive analysis has also been used to trigger aggressive internal improvement programs.[4]

Using strategic maps (Fig. 1-7), managers can understand industry dynamics, interrelationships among competitors, and the potential for exploitation of a competitive situation.

Still, all these tools were big-picture tools that did not provide enough detail for managers to answer that single gnawing question: *How* to improve competitive position? All the tools discussed thus far in this chapter are used by managers as part of overall competitive strategic analyses. These analyses look at one's competitors and the products or services they produce and are used as the basis for making big-picture strategic decisions. Should the company enter a new market? When? With which competitor should the company stand toe to toe and fight, and with which competitor should the company avoid confrontation at all costs? What does a competitor's shift in strategy mean? Should the company exit a particular market segment? What will happen if the company raises its prices? Lowers them?

---

[4]A. Steven Walleck, J. David O'Halloran, and Charles A. Leader, "Benchmarking World-Class Performance," *The McKinsey Quarterly*, no. 1, 1991.

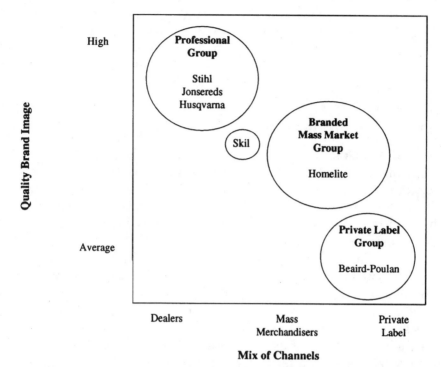

Figure 1-7. Illustrative map of the U.S. chain saw industry. (Reprinted with the permission of The Free Press, a Division of Macmillan, Inc. from *Competitive Strategy: Techniques for Analyzing Industries and Competitors* by Michael E. Porter. Copyright © 1980 by The Free Press.)

What these analyses did not do was stimulate improvement at an operational level, which is where the strategic decisions a company makes are executed and the fate of the strategy is decided. And these analyses were usually done by some staff person or group in an ivory tower, which meant that any analysis that did point to weakness in a company's own operations was often inherently suspect to the people on the line.

Best of all, the jury is out on whether corporate strategy can really be defined and, assuming it can, whether this is of any use. Once again, from *The Economist* (March 30, 1993, p. 76):

> More puzzling is the fact that the consultants and theorists jostling to advise businesses cannot even agree on the most

basic of all questions: what, precisely, is a corporate strategy?...A growing number of businessmen now question whether thinking consciously about an overall strategy is of any benefit at all to big firms. Grabbing opportunities or coping with blows as they arise may make more sense.

Enter Benchmarking

Andrall Pearson of Harvard Business School, former chief executive officer (CEO) of Pepsi and managing director at McKinsey & Company, once said, "I'll take solid execution over brilliant strategy any day." And Arthur Rock, venture capitalist who helped companies like Apple Computer, Intel, Fairchild Semiconductor, and Teledyne get started, said, "Strategy is easy. Tactics are hard."[5] Execution and tactics are where benchmarking becomes important. It was a logical, though perhaps not obvious, next step in the evolution of strategic planning and related tools. Benchmarking does not replace strategic planning, but supports it. Benchmarking takes strategic analysis to the next level of detail, which is necessary to win on the front lines. It looks at *how* a product or service is produced. It is not limited to competitors, either. Benchmarking can be used to study *any* company that may make a similar product or perform a similar process or activity, whether it is in the benchmarking team's industry or not. And benchmarking is usually done by a team that includes people from the line process being studied, which means that the line-versus-staff "we don't believe you" battles are never fought.[6]

Benchmarking has been growing more popular in the United States at least since Xerox began doing it in the late 1970s. A growing number of progressive companies began using benchmarking during the early 1980s, regardless of what they called

[5]Arthur Rock, "Strategy vs. Tactics from a Venture Capitalist," *Harvard Business Review*, November–December 1987, p. 63.

[6]See A. Steven Walleck, J. David O'Halloran, and Charles A. Leader, "Benchmarking World-Class Performance," *The McKinsey Quarterly*, no. 1, 1991, for further discussion of differences between competitive analysis and benchmarking. Many of the ideas in this paragraph were drawn from that work.

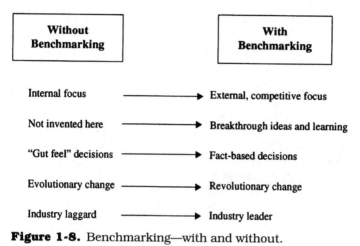

Figure 1-8. Benchmarking—with and without.

it, and a consulting firm or two seized the day by offering benchmarking services to companies that liked the idea but did not perform the practice in house.

According to Xerox, the benefits to be derived from benchmarking are quite obvious (see Fig. 1-8). The original framers of the Malcolm Baldrige National Quality Award included benchmarking among the criteria for the award, which has really moved it from the arcane to the best seller list of managerial tools available today. In the award's *1993 Applications Guidelines*, benchmarking appears in six of the seven Baldrige categories.[7] If benchmarking is a fad, it is fooling a lot of people, because some of the best and brightest in the United States are pushing it as a means of helping us regain our competitive edge.

---

[7]Category 2, Information and Analysis: Item 2.2, Competitive Comparisons and Benchmarks. Category 3, Strategic Quality Planning: Item 3.2, Quality and Performance Plans. Category 4, Human Resource Development and Management: Item 4.5, Employee Well-Being and Morale. Category 5, Management of Process Quality: Item 5.2, Process Management: Product and Service Production and Delivery Processes; Item 5.3, Process Management— Business Processes and Support Services. Category 6, Quality and Operational Results: Item 6.1, Product and Service Quality Results; Item 6.2, Company Operational Results; Item 6.3, Business Process and Support Service Results; Item 6.4, Supplier Quality Results. Category 7, Customer Focus and Satisfaction: Item 7.5, Customer Satisfaction Comparison.

David Garvin, a Harvard Business School professor and member of the original board of overseers of the Malcolm Baldrige National Quality Award, has this to say about benchmarking:

> In many ways, this spirit of cooperation is the legacy of the Baldrige Award. Winners are compelled by law to share their knowledge; that they have done so without suffering competitively has led other companies to follow suit. Benchmarking is by definition a cooperative activity, and it is an award requirement. Even warring factions of the quality movement have united under the Baldrige banner. To become more competitive, American companies have discovered cooperation.[8]

Enough about how benchmarking got here. The rest of this book will tell you what it is and how it works.

[8]David A. Garvin, "How the Baldrige Award Really Works," *Harvard Business Review*, November–December 1991, p. 93. Copyright © 1991 by the President and Fellows of Harvard College; all rights reserved.

047199

2

What Is Benchmarking?

Nam et ipsa scientia potestas est.
SIR FRANCIS BACON

Introduction

To be concise in introducing this chapter, benchmarking is two things: setting goals by using objective, external standards and learning from others—learning how much and, perhaps more important, learning how. One can add nuances or variations on the theme, but why bother? Simple is best. Benchmarking is not a numbers-only exercise. Setting quantitative goals, often called *metrics*, through benchmarking is arguably the best way to set goals, but keep in mind that setting goals comparable to or beyond those of the best-in-class without understanding the underlying processes that enable the best-in-class to achieve their results can be useless or worse. Understanding *how* the companies you study achieve their results is usually more important and valuable than obtaining some precisely quantified metrics. Keep this in mind during your benchmarking, and you will reduce the risk of losing sight of what you hope to get from a study—valuable learning.

SOUTH PUGET SOUND LIBRARY

A Tremendous Pool of Knowledge Exists

Benchmarking is not brain surgery. It is, plain and simple, learning from others. Identify them; study them; and improve based on what you have learned. When King Edward of England introduced the use of gunpowder in warfare against the French at Crecy, the days of the longbow were numbered. When Robert Fulton introduced steam power to the movement of ships, wind-powered vessels were soon relegated largely to leisure use in much of the world. The competition studied Edward and Fulton and brought their own performance up to a new level. We benchmark all the time in our daily lives, too. *Quality Progress* magazine ran a superb everyday example of benchmarking in a recent issue:

> Suppose that Mr. Puttputt, an avid golfer, has the opportunity to attend a golf workshop that features some of the best golfers in the world. He has the basics down but wants to improve what is currently giving him the biggest headache—putting. Mr. Puttputt knows that one of the world's best putters, Mr. Eagle, is going to be there, so he reads Mr. Eagle's book and watches Mr. Eagle's videotape in anticipation—he doesn't want to waste valuable time asking already answered questions. While reading the book and watching the videotape, Mr. Puttputt analyzes his own putting stroke in his mind (or in the living room, if no one is around).
>
> Finally, the day arrives. At the seminar, Mr. Puttputt meets Mr. Eagle, asks questions, and gets all the information he can (along with an autograph). On the way home, Mr. Puttputt mulls over all he has learned, deciding what will be most helpful in improving his putting. The very next morning he is on the green, applying what he learned in hope of becoming as good—or even better than—Mr. Eagle.[1]

This example is benchmarking in its essence, and it contains many of the subtle nuances of benchmarking to which you will be exposed throughout this book (e.g., gathering data on target companies before approaching them, analyzing your own activity, maximizing a site visit opportunity, recording your learning

---

[1]Karen Bemowski, "The Benchmarking Bandwagon," *Quality Progress*, January 1991, p. 19. Copyright © 1991, Quality Press, Milwaukee.

while it is fresh in your mind, and putting what you have learned to work as soon as possible).

A tremendous pool of knowledge and experience exists in the collective brains of the world's managers and workers. Knowledge about every facet of running an organization, from developing a new product to purchasing the inputs used to create the product or service item, to producing it, to developing and running the MIS, to providing managers with the information they need to make good decisions, to selling and servicing the product. Imagine a company—call it All-Star Manufacturing— that had the world's best "players" in R&D, purchasing, manufacturing, MIS, sales, service, and all the rest of the links in Porter's value chain.[2] Everything they did was best-in-class, and since they had the best infrastructure in the world, all groups worked together harmoniously to optimize the various processes.

A fantasy, this all-star team. And since the business world is so broad and dynamic, one would never be certain for long that the truly best players were the ones on this particular team anyway. But the very idea of such a team is only one step removed from the principle of benchmarking, which is to learn from the best. The aim of benchmarking is to tap into that tremendous pool of knowledge so that knowledge—the collective learning and experience of others—can be used by those who wish to improve their own organizations. Benchmarking is becoming so widely practiced for three primary reasons:

- *It is a more efficient way to make improvements.* Managers can eliminate the old process of trial-and-error learning, which often results in the reinvention of the wheel anyway. Managers can use processes that others have already proved effective and can concentrate their original tabula rasa thought on developing ways to improve these processes or tailor them to fit their own organizations' existing culture and processes.

- *It helps organizations make improvements faster.* Time has become such an important factor in competition today that managers in many industries are compelled to find ways to do things better faster. A mature benchmarking capability

[2]See Chap. 1 for a discussion of Porter's value chain.

within an organization will enable that organization to do things better and faster by working through the benchmarking process more quickly.

- *It has the potential to bring corporate America's collective performance up significantly.* Imagine that every company's after-sales service and support is as good as Xerox's. Imagine that every company's purchasing processes are as good as Cummins Engine's, or every company's warehousing and distribution is as good as L.L. Bean's. The United States would be much more productive if everyone benchmarked continually and brought key processes up to world-class standards. We could be faced with a large unemployment problem, however, if every organization improved its key processes through benchmarking.[3] Realistically and statistically, benchmarking will be like every other process, where the span of all practicing organizations' abilities to harness the power and resultant benefits of benchmarking can be placed neatly under a bell-shaped curve. Those who do it best will make great strides forward. Those who don't...well?

That's an overview of benchmarking. In practice, benchmarking is about giving one's organization a competitive advantage and outperforming the competition.

Figure 2-1 shows a benchmarking process that can be applied to nearly all business situations. The methodology is simple:

- Determine which value activities in your organization are the activities where improvement will allow the business to gain the most through benchmarking.

- Determine the key factors, or the drivers, of these value activities.

---

[3]George Pavlov, when manager of sales strategy at NeXT Computer, hypothesized that 25 percent of the U.S. workforce could be "redeployed" if management used benchmarking to set goals and provided remaining workers with the knowledge and wherewithal to meet those goals and measure-and-display charts to keep workers focused on the tasks at hand. The macroeconomic chaos one envisions in such a world would likely bring Karl Marx back from his grave.

Yet, efficiencies breed opportunities; automation was feared because it would eliminate jobs—and it did. It also generated new professions, new demands, and greater job opportunities.

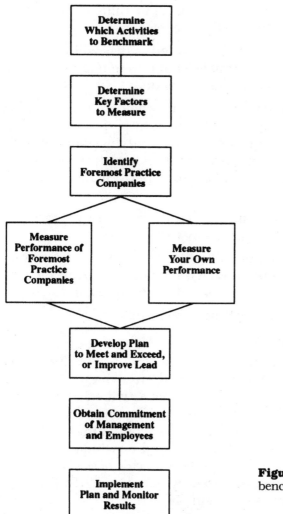

Figure 2-1. The eight-step benchmarking process.

- Identify the companies with foremost practices in these value activities. These foremost practices may be found at competitors or companies from unrelated industries—any companies that perform the value activities extremely well. For example, Xerox benchmarked L.L. Bean's warehousing and distribution system after determining that Bean's practices were the foremost practices in that particular activity. *Foremost-practice* companies are those companies that perform value activities at the lowest cost or the highest level of value to customers, whichever is appropriate to the circumstances.

- Measure the foremost practices in terms that will allow you not only to quantify performance but also to understand *why* and *how* they achieve the results they do.

- Measure your own performance and compare it to the best. You will likely have measured your own performance before measuring that of the firms you are studying. But you may find that you need to measure your own performance again because the best measure their own performance by using different measures.

- Develop plans to meet and exceed the foremost practices, or further your lead, as the case may be.

- Obtain commitment from all levels of the organization that are involved in the plan, which is made easier by the evidence provided by a sound benchmarking study.

- Implement the plan and monitor the results.

The applications for benchmarking are infinite. Assuming some finite amount of resources to devote to benchmarking projects, however, most organizations do well to establish some guidelines in determining what function, activities, or processes will be studied as part of their benchmarking programs.

There are many ways to practice benchmarking, and most organizations that have institutionalized benchmarking have tailored the basic process to meet their specific needs. For example, benchmarking processes used by Alcoa and AT&T follow, and one used by Xerox is shown in Fig. 2-2.

Alcoa's Six Steps to Benchmarking[4]

1. Deciding What to Benchmark. The project sponsor (the owner of the product, process, or service) identifies potential topics to benchmark. To check the relevancy and validity of a topic, the following questions are answered:

---

[4]Karen Bemowski, "The Benchmarking Bandwagon," *Quality Progress,* January 1991, pp. 22–23.

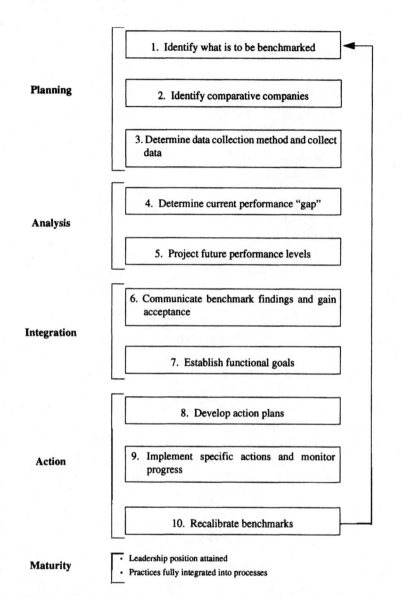

Figure 2-2. Benchmarking process—Robert Camp and Xerox. (Source: Robert C. Camp, *Benchmarking—The Search for Industry Best Practices That Lead to Superior Performance,* Quality Press, Milwaukee, 1989.)

- Is the topic important to the customers?
- Is the topic consistent with Alcoa's mission, values, and milestones?
- Does the topic reflect an important business need?
- Is the topic significant in terms of costs or key nonfinancial indicators?
- Is the topic an area where additional information could influence plans and actions?

The output from this step is a purpose statement that describes the topic to be benchmarked and guides the activities of the benchmarking team.

Benchmarking topics are also selected when teams working through Alcoa's eight-step quality improvement process find themselves asking, "Has anyone ever faced a similar problem? What was done about it?"

2. Planning the Benchmarking Project. A team leader is chosen. The project sponsor is the ideal team leader, but if that person cannot assume this role, she or he can designate the team leader. The leader, who will be responsible for seeing that the project is successfully completed, should have the authority to make changes in processes, products, and services based on benchmarking information.

Next, the team members are selected, based on the range of skills needed for the benchmarking project. The team's first task is to refine the benchmarking purpose statement by answering these questions:

- Who are the customers for the study?
- What is the scope of the study?
- What characteristics will be measured?
- What information about the topic is readily available?

Finally, the team submits to the sponsor a project proposal that includes all the information obtained up to this point. Once the project sponsor approves the proposal, the team moves to step 3.

3. Understanding Your Own Performance. This step is devoted to self-study. The team examines the factors that influence performance to learn which characteristics are most important and which least important. The team also learns what data relate to the important characteristics and how to collect and measure those data. This process itself might reveal new ways to overcome specific barriers. The collected performance data are used to create the baseline and structure for benchmarking comparisons.

4. Studying Others. In this step, the team

- Identifies benchmarking candidates
- Narrows the list to a few candidates
- Prepares general and specific questions
- Decides the best way to get those questions answered
- Performs the study

Also included in this step are some guidelines regarding ethical and legal issues that might arise during the study.

5. Learning from the Data. The team analyzes the data collected, quantifies performance gaps, and identifies which pieces of information might be particularly useful for improving performance.

6. Using the Findings. The team works with the project sponsor to determine how the benchmarking findings can be best used and what other organizations in the company can benefit from this work.

An AT&T 12-Step Benchmarking Process[5]

The benchmarking process used by AT&T's Material Management Services Division has 12 steps, which are divided

[5]Ibid., pp. 20–21.

into two distinct categories. Steps 1 through 6 are referred to as the *first-things-first* steps, because they help prevent barriers that could hinder or even destroy the benchmarking process. Steps 7 through 12 are called *process steps*, because they outline the process by which benchmarking is carried out.

1. Determine Who the Clients Are. The clients—the people who will use the benchmarking information to improve their processes—vary depending on their companies' organizational structures. AT&T has two sets of clients: *process owners and planners,* who are responsible for the continued improvement in their divisions, and *business units,* which are the divisions' end customers.

2. Advance the Clients from the Literacy Stage to the Champion Stage. Clients are taken beyond just knowing what benchmarking is to visualizing how it will help them develop best practices. This helps develop client support and patience.

3. Test the Environment. Time is spent with clients to determine the extent of their buy-in and commitment of resources, resulting in realistic expectations and exposing barriers.

4. Determine Urgency. The sense of urgency within the client's environment is determined to see what degree of optimism there is for the project. Panic and a so-what attitude are formidable barriers to successful benchmarking. In these cases, benchmarking degenerates to the level of tours, and little objective fact finding or self-appraisal is done. The most conducive situations for benchmarking occur when the client is reengineering processes or striving to become the best in class. The middle ground is made up of clients who want to evaluate themselves or continue to improve in their areas.

5. Determine the Scope and Type of Benchmarking Needed. The scope and type of benchmarking needed depend on the clients' sense of urgency, their environment, and their understanding of, and willingness to commit to, the benchmarking process.

The scope—the required time, people, resources—is proportional to the potential payback. For example, benchmarking as a simple task takes a lot less time and workforce than benchmarking as an operations process, but the payback will also be less. AT&T divides scope into four levels, which determine the complexity and amount of time needed to implement the benchmarking project, as well as the potential payback.

The type of benchmarking determines what type of organization will be benchmarked. AT&T benchmarks three types of organizations: best-in-class performers regardless of industry and location, internal-best performers, and competitive-best performers.

6. Select and Prepare the Team. By working with the client, a benchmarking team of six to eight members is selected. The team is responsible for putting together the benchmarking proposal and integrating the resultant recommended actions into the business plan for implementation. However, only two to four team members actually visit the organization being benchmarked. All team members are fully trained in the benchmarking process.

7. Overlay the Benchmarking Process onto the Business Planning Process. The team reaffirms that benchmarking is accepted by upper management as part of the business planning process.

8. Develop the Benchmarking Plan. The degree of organization and teamwork developed before the visit affects the team's effectiveness. When developing the benchmarking plan, the team

- Prepares a mission statement that formalizes expectations.
- Prepares for data collection. Roles are assigned to team members, and the subject being benchmarked is thoroughly analyzed. The team identifies work processes to be studied and develops critical, open-ended questions.
- Develops a profile for selecting the benchmark partners (i.e., the organization being benchmarked). The team determines

which benchmark performance characteristics are most important to the client's interest, to the specific problems and work processes identified for benchmarking, and to any performance aspects of special importance (e.g., key measurements, similar volumes of activity, customer bases, product types, sales channels).

- Does research. The team spends considerable time at a library or research center, learning about the organizations being considered for benchmarking. This research is used not only to select an appropriate benchmark partner, but also to prepare properly for the site visit once the partner is chosen. This research does not, however, replace the actual visit.

- Develops scripts. The team develops written scripts to help organize and manage the site visit. These scripts include both the in-depth analyses of the functions, processes, tasks, etc. identified for benchmarking and the open-ended interview questions.

- Describes present operations. The team describes how the client performs the function being benchmarked. Answering the script questions usually produces a solid description of current operations.

- Indicates metrics. A well-documented set of metrics is included in the present operations description. Comparisons of metrics are used to select the benchmark partner and to understand that organization's performance.

- Sets up visits and protocol. With the planning done and the prospective benchmark partner chosen, the team contacts the prospective benchmark partner. A letter requesting a visit is followed by a phone call, to discuss the intent in greater detail and to assess the prospects of benchmarking this organization. Once the organization agrees to be a benchmarking partner, it is sent the script to help it prepare for the visit. The sales, service, and other departments within the client organization are told who the benchmarking partner is since much can be learned from these departments and their protocol.

The visit is then made. The team understands that it is participating in an intelligence patrol, not a combat patrol. The team does not defend how its client performs the function being benchmarked. The team makes sure that each question is answered and documented.

9. Analyze the Data. The information gathered is compared to the client's present operation to determine where improvements can be made. Where possible, findings and opportunities are quantified. To ensure a successful analysis, the team

- Organizes its visit documentation into flowcharts, narratives, matrix formats, comparison charts, etc., to clearly summarize the findings. This is done for each visit. The results are then integrated into one analysis summary, which is distilled into opportunities for quantification.
- Makes sure the present operations description is accurate and effectively compared to the benchmark findings.
- Makes sure that sound quality principles are followed.
- Avoids follow-up visits to the benchmark partner (although revisits might be necessary to collect detailed data).
- Identifies opportunities for improvement.

10. Integrate the Recommended Actions. The client takes the team's recommended actions and integrates them into the planning, budgeting, financial, service, and other applicable processes. How this is done depends on the client's vehicles for implementing and tracking change and service costs. The recommended actions are accounted for in the client's budget.

11. Take Action. The client implements the action outlined in the various planning processes. The normal procedures for implementing change are followed. Ownership is assigned, and progress is tracked for the improvement process.

12. Continue Improvement. Once the opportunities are seized, the client makes sure that continuous improvement activities are in place by institutionalizing benchmarking in its planning and continuous-improvement processes. Benchmarks are periodically recalibrated because they change as new leaders emerge.

Although these four benchmarking processes differ somewhat, the differences are largely semantic. Most of the benchmarking that is taking place across the country today—indeed, across the world—follows a roughly similar methodology to identify, study, learn, and improve. And most of it falls into three categories: competitive benchmarking, cooperative benchmarking, and collaborative benchmarking.

Three Common Types of Benchmarking

Competitive Benchmarking

Competitive benchmarking is the most difficult form of benchmarking because, as its name suggests, target companies are not usually interested in helping the benchmarking team. Data gathering, which almost always is the most time-consuming task in any benchmarking endeavor, is made much more difficult when your targets are your competitors. They do not want you to study them and may go to great lengths to thwart your efforts (e.g., through gag orders) if they learn you are doing so.

Competitive benchmarking means measuring your functions, processes, activities, products, or services against those of your competitors and improving yours so that they are, ideally, the best-in-class but, at a minimum, better than those of your competitors. What do you want to know about your competition? Make a wish list. Chances are, you can learn outright many of the things on your list, and you can estimate with reasonable precision most of the rest. And you can do it legally and ethically. *Nam et ipsa scientia potestas est*—Knowledge is power, but only if you use it. Collecting and analyzing data about your competitors is a fairly straightforward, though time-consuming, process. Most benchmarking studies that fail do so not because data could not be collected but because managers do not use the information they have spent good resources to understand to effect positive changes based on what they have learned. A

Competitive Benchmarking	
Your Organization	Your Competitors
• *What* are you doing. • *How* you are doing. • *How well* you are doing it.	• *What* are they doing. • *How* they are doing. • *How well* they are doing it.
Result: Increased awareness of your organization.	Result: Increased awareness of your competitors.

Figure 2-3. Competitive benchmarking. (Source: Xerox Corp.)

synoptic view of competitive benchmarking as practiced at Xerox is shown in Fig. 2-3.

Cooperative Benchmarking

Cooperative and collaborative benchmarking (discussed next) seem to be the most talked-about forms of benchmarking today because they are relatively easy to practice, because they make interesting news copy, and because teams engaged in competitive benchmarking usually keep their mouths shut about it. In cooperative benchmarking, an organization that desires to improve a particular activity through benchmarking contacts best-in-class[6] firms and asks them if they will be willing to share knowledge with the benchmarking team. The target companies are usually not direct competitors of the benchmarking company, which is a key factor in securing cooperation. For example,

- Managers at a west coast-based electric utility benchmarked the operating performance of their coal-fired generating station against similar plants in other regions of the country. Although competition is being introduced to the electric utili-

---

[6]In a strict grammatical sense, there can only be one best-in-class firm in a given activity. In a benchmarking sense, however, the term *best in class* is often used interchangeably with such terms as *world class, foremost practice,* and plain old *a lot better than us.* Some tolerance for ambiguity is a virtue in using any of these terms. The principle of benchmarking requires study of organizations that perform a specific activity better than your organization does and, of course, the closer they are to being the "best," the better the learning that can be potentially obtained from studying them.

ty industry, managers at the selected target plants did not consider the benchmarking plant a direct competitor. As a result, the managers were most cooperative, which allowed the study to pay off very well for the benchmarking team.

- Managers at a CAD/CAM software company benchmarked Hewlett Packard, IBM, and Xerox in the area of after-sales service and support. Without competitive considerations to consider, managers of the three target companies were able and willing to share tremendous knowledge with the benchmarking team.

In cooperative benchmarking, the knowledge usually flows in one direction—from the target companies to the benchmarking team. Although the benchmarking team often offers to provide the target companies with some benefit in return, the targets typically give more than they receive.

Collaborative Benchmarking

In collaborative benchmarking, a group of firms share knowledge about a particular activity, all hoping to improve based upon what they learn. Sometimes, a third party often serves as coordinator, collector, and distributor of data, although an increasing number of firms are managing their own collaborative studies. For example,

- Managers from the training organization of a large financial services firm organized an ad hoc consortium to study training processes at a number of U.S. firms that are considered leaders in training. Participants included American Airlines, AT&T, NCR, Quad Graphics, Solectron, and USAA, all of whom openly shared detailed data about their training processes with one another.
- Managers from one of the Malcolm Baldrige National Quality Award winners convened an ad hoc consortium to study the use of customer satisfaction data at a number of U.S. companies that are considered leaders in this area. Participants in the study included five Baldrige winners—AT&T, IBM-Rochester, Motorola, Solectron, and Zytec—Time-Life, and MBNA, the financial services company.

And there are many other collaborative efforts that have produced great results. One consortium that includes Arthur Andersen, Eastman-Kodak, and Xerox, among others, has been sharing data and knowledge in a number of areas, with many positive gains to show for it. And the American Productivity and Quality Center (APQC) in Houston is compiling a benchmarking clearinghouse database, which they claim will be a database of "best practices in both abstract and full text form."[7] Reviews on the value added of the APQC's clearinghouse have been mixed, as of this writing. You should conduct your own due diligence before signing on as a member of any benchmarking database or clearinghouse to ensure that you will get what you expect for your money.

Remember that not all benchmarking is created equal. Some collaborative efforts, although called benchmarking, are typically just data-sharing exercises that address the question "How much?" but fall short on answering the question "How?" In that sense, these studies are not benchmarking as it is strictly defined, in which the benchmarking team learns not just how much improvement can be made but also how to make it. These exercises are frequently orchestrated by a third party who, for an additional fee, will come in and help participants answer the question "How?" Of course, they may be doing this for all the participants...*caveat emptor.*

Internal Benchmarking

Internal benchmarking is a form of collaborative benchmarking that many large organizations use to identify best in-house practices and disseminate the knowledge about those practices to other groups in the organization. Internal benchmarking is frequently done by larger companies as the first step in what will eventually become an outward-focused study. There are a couple of reasons for this. First, it enables the benchmarking team to climb the learning curve, i.e., develop or enhance its fundamental base of knowledge about the issue being studied, with help from colleagues, who should have fewer reservations

[7]APQC, Houston, Texas.

about sharing than counterparts at other companies. Second, it provides the benchmarking team with more to offer to managers at external target companies when approaching them about cooperating or collaborating in a study. The ability to bring some value to the table in benchmarking, and to make the flow of knowledge two-way instead of one-way, is becoming increasingly valuable as more companies begin practicing benchmarking. Chapter 6 discusses this concept in greater detail under the heading "Identify Your Prisoners."

Look Out

In summary, what all three benchmarking approaches do is push managers to look outside their organization at their competitors or other best-in-class companies and use the collective knowledge of these other organizations to make their own organization stronger. An external focus in seeking improvement can help organizations make quantum leaps in performance instead of incremental improvements like those that are often made when the improvement process is merely tinkering with an activity that requires a major overhaul. Quantum leaps instead of incremental improvements. Think about it.

3
Fire!

Live every day like your hair's on fire.
G. A. PAVLOV

Pardon the title of this chapter, but I had to grab your attention somehow. For the sake of those of you who did not read Chap. 2 or did but want to hear it again, *benchmarking is not a numbers-only exercise*. Numbers, or *metrics* as they are frequently called, are at best only half the benchmarking process. Understanding the managerial and operational processes that enable best-in-class organizations to achieve the results—the "numbers"—that they do is usually more valuable than quantifying precisely the results themselves. You have not finished benchmarking when you have quantified your benchmark. You must understand the underlying processes for your benchmarking study to be of real value. Only this understanding is going to allow you to make improvements in your own activities based on what you have learned.

Also, benchmarking usually is not glamorous. The benchmarking stories you read—company X managers going to Disneyland to benchmark the facilities management process or company Y managers going to the Indianapolis 500 to watch pit crews in operation, hopefully seeking some insight into production line changeovers—make great news copy, which is half the reason they are used to describe benchmarking studies. But most benchmarking studies do not make great copy. The truth is that most benchmarking studies are not glamorous—they are rigorous, hard work, and discussing them at cocktail parties is not going to make you the center of attention or add to your reputation as a scintillating wit. Take my word for it.

4

Why Benchmark?

If you're not doing it, you probably should be.   ANDRALL PEARSON

Introduction

The most widely heralded benchmarking success story is probably that of Xerox Corporation in the1980s. But for every Xerox story there are literally hundreds of other examples of benchmarking success that U.S. managers are living every day. Almost all of the *Fortune* 500 firms are benchmarking, and based on comments from managers at many of these companies, the process is achieving the desired result, which is increased competitiveness for their companies.

More than just the *Fortune* 500 companies are benchmarking, however. The MIT Commission on Industrial Productivity in 1989 reported that "a characteristic of all the best-practice American firms, large or small, is an emphasis on competitive benchmarking: Comparing the performance of their products and work processes with those of World Leaders in order to achieve improvement and measure progress."[1] And the inclusion of benchmarking among the application guidelines of the

---

[1]Dertouzos, Michael L., Lester, Richard K., and Solow, Robert M., *Made in America*, MIT Press, Cambridge, MA, 1989. Reprinted with permission.

Malcolm Baldrige National Quality Award makes benchmarking a requirement for all companies that are serious about improving their competitiveness through quality.

So what? Who cares what the *Fortune* 500 are doing? And who cares about the Baldrige Award? Why should you benchmark? Because it makes so much sense, that's why. Identify the best; study and learn from them; and implement improvements that will work in your organization based on that learning. It is a simple, compelling idea. Applying the logic of two heads being better than one to solving business problems, the accumulated knowledge of two, three, four, or a half-dozen organizations to solve a particular issue of one must be better than that single organization solving it alone. Other organizations have devoted brain power to solving their own problems. Why not tap into that resource and enable the benefits of that learning to be realized more than once? Assuming appropriate resolution of any potential legal issues, that learning *should* be used more than once. It makes so much sense for managers across the nation to learn from one another when possible. The accumulated learning of corporate U.S. managers is a vast resource that is just beginning to be recognized and tapped. Benchmarking is starting to tap that resource and share the wealth—not in a manner that is going to reduce competition but in a manner that may bring back some of the collective competitiveness that corporate America seems to have lost.

The Xerox Story

Take a look at Fig. 4-1, which shows annual changes in revenue and pretax income at Xerox from 1961 to 1982. Scrutinize the two lines closely to see if you can discern a trend. You don't need advanced analytical skills to identify it, do you? Xerox of the 1960s and 1970s was a very different company from the organization that leads the copier industry today. The company that brought the xerography process to the market and simplified our lives (remember mimeograph machines?) enjoyed many years of near-monopoly status in the copier industry. Patent protection and the luxury of extremely high growth year after year combined to the unenviable result of a company that was fat, dumb, happy, and arrogant—not a good mix.

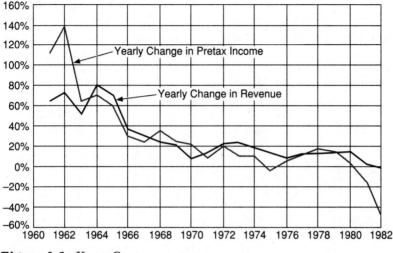

Figure 4-1. Xerox Corp.

Excerpts from the chairman's letter to shareholders in Xerox's annual reports are illustrative of the complacency that had set in. Let's sample a few. In the interest of fairness, I will admit to having chosen selectively those passages that, in hindsight, will elicit some chuckles at Xerox's expense. But I did not have to search too hard, and Xerox people today are the first to admit that the old Xerox almost got what it had coming. So on with it.

From Xerox's 1969 Annual Report:

> On September 9, 1969, Xerox officials in Rochester lit the candles on a 12-foot cake to celebrate the 10th anniversary of our first major xerographic product.
>
> The 914 copier—which some believed to have a very small market—launched us into a decade of growth that has few precedents in American business.

From Xerox's 1972 Annual Report:

> At the risk of becoming too philosophical here, let me simply say that when any system becomes so complex and so controlled that no single person is free to make or to contribute to a meaningful decision...then that system is ultimately destined for extinction.
>
> Xerox is by no means near that point....

FTC Files a Complaint.   This January, the Federal Trade Commission filed a complaint against Xerox, alleging that we have monopolized the copier market and hindered effective competition. Earlier, the FTC had announced its intention to file and had given Xerox the choice of negotiating a consent settlement. We chose not to do so, a decision not difficult to make.

Among the actions the FTC wishes us to take are...to offer unrestricted licensing of all Xerox copier patents and related know-how, now and for the next 20 years.

Charges Are without Foundation.   We quickly set forth our position that we consider the Commission's case to be without merit and its charges ill-conceived.

Moreover, given the fact that there are already some 50 competitors in the U.S. market alone, along with an unknown number in the rest of the world, many of whose plans must certainly include entry into the United States, the Commission's charges are in our view without foundation, either in fact or in theory.

A decision not difficult to make. Don't sell those shares, shareholders. We know what we are doing. Um, well, but we don't know how many copier companies are in the rest of the world....

From the 1975 Annual Report:

For the first time, Xerox worldwide revenues exceeded $4 billion. Nevertheless, 1975 was a difficult year and the first time since 1951 that our earnings declined.

Copy Volume High.   We were encouraged that copy volume, worldwide, on our equipment remained strong in 1975. Machine placements, however, were soft, particularly in the Untied States due to the economic environment and increased competition in the copier/duplicator market.

FTC Settlement.   On July 29, the U.S. Federal Trade Commission issued a consent order settling its antitrust complaint against Xerox. It required us to license our copier/duplicator patents; also, to make certain know-how available to competitors manufacturing in the United States.

Reductions in Work Force.   Since ours is a people-intensive business, the sharp decline in new machine placements...regrettably required us to reduce the work force. Xerox had to release more than 5,000 employees during the past calendar year.

From the 1977 Annual Report:

Xerox at this moment is a company in significant transition in its technologies, its products and its relation to its markets.

We are changing of necessity...

We are changing by design...

And we are changing by choice...

By choice. Someone is hitting you in the face with a shovel. If you ask them to stop, they will. Should you ask or not? Your choice...

The fact was that Xerox, in 1975, was still the undisputed leader in the copier industry. It was estimated that Xerox's share of market was greater than 75 percent, and over the prior 5 years, Xerox's revenues had been growing at an annual rate in excess of 25 percent. But change was coming. In Xerox's own words:

> Two dramatic changes occurred during the late '70s: the copier industry as a whole was targeted by Japanese firms, and Federal Trade Commission settlements required Xerox to open international access to key patents. Aggressive Japanese companies attacked the low end of the market with small, high-quality and low-priced copiers. The Japanese then built upon this success to penetrate the mid-range market, while IBM and Eastman Kodak introduced competing high-end equipment. Xerox's period of unchallenged market dominance was over.
>
> This experience is not dissimilar to that of many other industries in the United States over the past decade.[2]

By the beginning of the 1980s, Xerox's market share had dropped more than 50 percent. The future looked very uncertain. It was a time for decisive action, but companies do not always recognize such points. Failure to act appropriately could result in Xerox's becoming noncompetitive or worse, much as Detroit did for a while under similar circumstances in the early 1980s when the Japanese targeted the automobile industry.

In 1979, though, Xerox had begun competitive benchmarking in a few of its operating units, and the power and value of benchmarking were beginning to be understood by Xerox managers. By 1981, benchmarking was being used throughout

---

[2]*The Xerox Quest for Quality and the National Quality Award,* Xerox Publication 700P91456.

Xerox. David Kearns, Xerox CEO, described the corporate effort as follows:

> We took competitive analysis one step further and came up with what we now call competitive benchmarking. It's an intense, in-depth study of what we think is our best competition. It's a continuing, never-ending process, and it's an integral part of our new and stronger emphasis on quality. Every department at Xerox should be benchmarking itself against its counterpart department at the best companies we compete with. We look at how they make a product...How much it costs them to make it...How they distribute it, market it, sell it and support it...How their organization works...What kind of technology they have. Then, we all go back and figure out what it takes to be better than they are in each of those areas.[3]

Benchmarking was a key component in Xerox's corporate betterment thrust, which was termed *Leadership through Quality*. By 1983, the Xerox turnaround had begun. Managers across the company were benchmarking best-in-class companies wherever the companies could be found. Included among the companies studied by Xerox were those shown in Fig. 4-2.

Over the next 5 years, Xerox underwent a massive culture shock as senior management made painfully obvious its intolerance of the status quo. Xerox managers have called this the *internal revolution*, led by David Kearns and Paul Allaire.

The result of Xerox's corporate effort, to which benchmarking contributed significantly, is well known. In 1989, Xerox Business Products and Systems won the Malcolm Baldrige National Quality Award and had regained much of its lost market share. All this had been accomplished in the face of ever-growing industry competition—by 1990 there were more than 100 companies making competing copying machines and dozens of companies competing in the office products and systems markets.

Figure 4-3 shows some of the highlights from Xerox's Baldrige application.

[3]*Competitive Benchmarking: The Path to a Leadership Position*, Xerox Publication 700P90261.

Company	Process
American Express	Collections
American Hospital Supply	Inventory control
AT&T	Research & development
Baxter International	Employee recognition and human resources management
Cummins Engine	Plant layout and design; supplier certification
Dow Chemical	Supplier certification
Florida Power & Light	The quality process
Hewlett-Packard	Research & development; engineering
L.L. Bean	Inventory control; distribution; telephonics
Marriott	Customer survey techniques
Milliken	Employee recognition
USAA	Telephonics

Note that none of these companies is a copier manufacturer. Had Xerox benchmarked only its competition, it would have failed to find significant opportunities to improve.

Figure 4-2. Selected Xerox benchmarking partners.

The Xerox story is well known, and Xerox managers are justifiably proud to spread it. Xerox's belief in benchmarking and the obvious benefits that benchmarking has brought to Xerox have helped managers throughout the United States learn how valuable benchmarking can be in making improvements in any business process. Those who are benchmarking rigorously are realizing similar improvements in their own processes. Those who are not might be well advised to pay heed to David Kearns' words on benchmarking: "Where companies go wrong is that they don't start benchmarking *before* they're threatened."

The AT&T Universal Card Services Story

An example of benchmarking used under an entirely different set of circumstances is AT&T Universal Card Services, the Visa- and MasterCard-issuing subsidiary of one of the world's best known companies and winner of a 1992 Malcolm Baldrige National Quality Award.

During Oscar night in 1990, AT&T introduced the Universal Card to the United States by kicking off a well-publicized

---

Highlights from Xerox's Malcolm Baldrige National Quality Award Application

Productivity

- Service visits per day have been increasing 4% per year.
- Product performance during the first 30 days of installation has increased 40%.
- Manufacturing lead times have been reduced 50%.
- Manufacturing labor and material overhead rates have been improved by 31% and 46% respectively.
- Customer retention rate is 20% better than U.S. industry average and Xerox is gaining customers at a rate of more than four new customers for every three customers lost.

People

- 75% of all Xerox employees are actively involved with quality-improvement or problem-solving projects.
- 94% of Xerox employees acknowledge that customer satisfaction is their top priority.
- Employee turnover is 17% better than the average reported by the Bureau of National Affairs.

Customer Processes

- Highly satisfied customers have increased 38% and 39% for copier/duplicator and printing systems respectively.
- Customer complaints to the President's Office decreased 60% and continue to decline.
- Customer satisfaction within Xerox Sales processes have improved 40%; Service processes, 18%; and Administrative processes, 21%.
- Billing quality has improved 35%.
- Service response time has improved 27%.
- Supply order returns have improved 38%.

Safety

- Product safety has improved 70% with an associated 90% decrease in claims. Xerox has not had a product liability judgment in the last five years.
- Xerox employees are three times safer on the job than around the home. There has never been an industrial fatality nor a major OSHA citation in Xerox.

Figure 4-3. Highlights from Xerox's Malcolm Baldrige National Quality Award application.

advertising campaign. No annual fees for life, lower interest rates, a discount on AT&T long-distance calls, and world-class service were just a few of the features offered to attract customers to the card. And attract customers they did. Within 78 days AT&T Universal Card Services had broken an industry record by signing up their one millionth account. By January 1992, they reached fifth place in the *Nilson Report's*[4] ranking of top general-purpose credit cards by charge volume and by number of cards outstanding, which exceeded 12 million.

[4]*Nilson Report* is one of the best-regarded credit card industry newsletters.

These were impressive results, and not at all due to luck. The Universal Card Services organization was built through benchmarking in a methodical and strategically sound way. Many of the best minds in the business were hired from industry competitors, and with them came, collectively, a very clear picture of what it would take, both strategically and operationally, to be a world-class competitor in the credit card industry. No annual fee, lower interest rates, long-distance discounts, and world-class customer service are just a few of the visible results of the benchmarking that built Universal Card Services. Supporting it all are world-class credit procedures, statementing, remittance processing, workstation design, and quality monitoring, to name a few of the areas in which Universal Card Services is world-class thanks to the knowledge of the industry's best that built the company. This was an extremely different set of circumstances from that which faced Xerox, but the end result was remarkably similar—a world-class competitor in its industry.

Read On

The next several chapters will help you

- Understand some of the arguments against benchmarking
- Plan a benchmarking study
- Execute the study
- Make improvements based on the study findings
- Implement a benchmarking process in your organization
- Understand where benchmarking is headed

Keep in mind as you read, especially if you are planning a study of your own, that the primary reason to use benchmarking should be as a means to build a sustainable competitive advantage for your organization. Benchmarking has diminished value if its end goal is something less.

5

Three Common Criticisms of Benchmarking

*It is better to be making the news than
taking it; to be an actor rather than a
critic.*   WINSTON S. CHURCHILL

Introduction

I thought I would get this chapter out of the way early.
Intellectual honesty makes it de rigueur, but I do not agree
largely with the criticisms summarized below.

Not everyone thinks benchmarking is a good thing. When
practiced poorly, it probably does more harm than good.
Managers at one of the Bell operating companies were continually seeking "the number." "Let's call so-and-so at Pacific Bell
and find out what his average customer complaint resolution
time is," was what they thought benchmarking was before they
became educated. Using "benchmarking" to set goals without
providing those who must meet those goals with an understanding of the underlying processes can cause a great deal of

frustration in the ranks. And burning up corporate resources on shoddy benchmarking is a terrible waste. But neither of these scenarios implies that benchmarking is not a powerful tool when used correctly.

There are, however, some common criticisms of benchmarking. Each is covered briefly below. Awareness of these criticisms can only help your benchmarking, especially if it helps you avoid some of the pitfalls that have been identified.

Spying

Some U.S. managers are reluctant to participate in what they term corporate *spying*. While shopping a few years ago, I met an old colleague and his wife. She asked where I was working and what I was doing. I told her that I was working with a firm that specialized in benchmarking, and I began to explain what that was when her husband broke in with two words: *industrial espionage*. Some people really believe that.

In Japan, knowing the competition is part of every manager's job description. This has been a contributing factor to the Japanese rise to a position of dominance in industries such as motorcycles, automobiles, consumer electronics, and many others. An entire industry of benchmarking firms in Japan gathers detailed competitor information for clients, but they do not have a catchy name for it like *benchmarking*. They just do it. The fact that the Japanese do it does not make those who feel competitive benchmarking is spying any less vehement in their arguments, however. Philosophies about issues such as this, which some feel are purely ethical in nature, don't change easily.

Copycatting

Another common criticism is that benchmarking results in *copycatting*, which reduces creativity and may be detrimental in the long run. This combined with aiming at targets as if they were fixed points to hit may, in fact, reduce the value of any management insight gleaned from the process. However, benchmarking is not supposed to make managers copycats. It is supposed to make them learn new ways to think about old problems. And any well-done benchmarking endeavor recognizes that competitors are moving targets and requires that managers adjust planned actions accordingly.[1]

Having the awareness that copycatting is a trap that one can fall into when benchmarking should help managers avoid it. Benchmarking is not about copying, it is about learning; and the difference between the two terms should be considered real, not just semantics.

Not Invented Here

Also holding benchmarking back in some organizations and, to some degree, in entire industries is the "not invented here" response some managers have for any learning that originates outside their organization. There also may be a fear of exposing an egregious organizational fault by comparing it to world-class standards, which are often discounted due to organizational differences anyway ("That way won't work for us—we're different"). This ostrichlike argument can be detrimental to an organization's health. Perhaps if managers, who cite "not invented here" as an argument against benchmarking, knew the Xerox story, they would be less resistant.

One still sees manifestations of the not-invented-here syndrome frequently accompanying the smartest and the youngest firms, especially when they have met with nothing but success during their short lives. And the "if it ain't broke, don't fix it" philosophy might apply as long as things are going well, but continuous improvement seems to be the new law of the universe. There is a lot of learning that can come from outside

---

[1]For a discussion of the dangers of copycatting and shooting at targets as if they were static, see Mark Whatley and Howard B. Aaron, "Companies That Target World Class Are Destined to Be Second Rate," *Quality Engineering*, 3(2), 207–213 (1990–1991).

one's organization. The sooner the not-invented-here organizations realize this, the better. In the late 1980s, a big splash was made by a new entrant to the computer hardware industry. Management claimed to be building "the next generation of computers," and on the team was a quality manager who tried to build a companywide preference for benchmarking. It never caught on. "Not invented here" was a way of life among the managers and employees of the company. They were the best and the brightest in the industry, if not the world, and were the first to make sure that fact was clear to all within earshot.

They are no longer in the hardware business. Those who remain, fewer than half of those who worked at the company the day before the announcement to leave the hardware business was made, will give it another try in the software business. Not invented here—not a good reason to eschew benchmarking.

6

Planning Your Benchmarking Study

An ounce of prevention...
   B. CONRAD NELSON

Introduction

Before you begin this chapter, consider the normal bell-shaped curve at the bottom of this page. Now think ahead to 5 years from now. If you were to look back on 5 years of benchmarking you and your colleagues had performed to that point and evaluate how well you did it, where would you want to be located

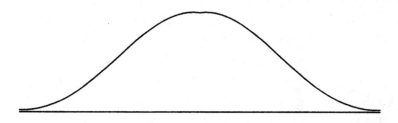

under this bell-shaped curve? Somewhere right of center, correct? And preferably to the far right.

Well, like so many "populations" in life, the population of all benchmarking that will be done over the next 5 years is going to fit rather nicely under a normal bell-shaped curve. And educating all the benchmarking practitioners is not necessarily going to skew the curve to the right, because *the deciding factor will be how committed to doing it well people are when they do it.* So *learn* how to do it right, and then do it as if your organization's life depends on it, because it just might. Are you going to run with the big dogs or sit on the porch?

Get It Right the First Time!

Each of the three phases of benchmarking—planning, execution, and implementation of benchmarking-based improvements—requires different skills to be done successfully. The planning phase (see Fig. 6-1) requires skill in analyzing the issues one chooses to address through benchmarking and then requires great organizational skills to ensure that the study is planned to be executed smoothly and successfully. Poor planning can make your benchmarking project a waste of time and resources, but many neophyte benchmarkers underestimate the importance of planning because of the relatively minor amount of time it represents as a component of the whole process. You should be patient and plan well the journey before embarking. The time you spend planning the study is a small fraction of the total time you will spend, but it is extremely valuable. *Don't dive into data collection and analysis!*

There are many opportunities for things to go wrong during a benchmarking study. Good planning can obviate many of them, but you can make mistakes in planning, too. These are the three most common mistakes in planning a benchmarking study:

- *Planning to benchmark the wrong activities.* This is fundamental. Understanding precisely the goal of your study, i.e., what you and your colleagues want benchmarking to do for you, is

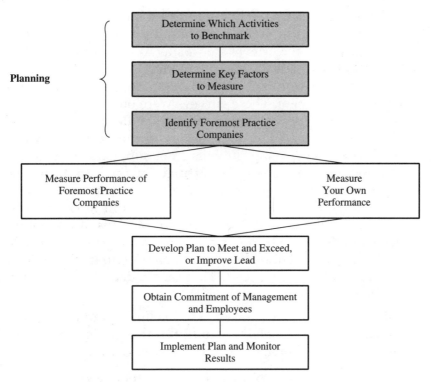

Figure 6-1. Benchmarking, the eight-step process.

critical to the success of the project. You need a vision. If you feel compelled to dive into a benchmarking study and spend many hours and dollars in collecting and analyzing data before developing this vision, squelch the urge by picturing how you will feel after you have pulled together your analysis, written your report, and presented it to your board of directors, whose collective reaction is "So what?"

You should benchmark activities where improvement is going to contribute significantly to the competitiveness of your organization. This is discussed further in this chapter and throughout this book.

- *Measuring something other than the key factors.* Managers at a wholesale building materials chain benchmarked some of the best-in-class retailers to develop an understanding of some of the key metrics in retailing that would allow them to improve customer satisfaction. They measured such things as customer/salesperson ratios, on-time delivery, and product availability, and they developed plans to meet or exceed the leaders in each of these measures. Meet or exceed they did, but the responses to subsequent customer satisfaction surveys surprised them. These surveys were often completed by the contractors' wives, who typically handle the administrative duties for their husbands. Many of their opinions were developed when paying the bills from suppliers such as the building materials chain. Unfortunately, the bills from the building materials chain were often replete with errors. The contractors' wives spent hours and hours in most months resolving billing errors. They were not big fans of the building materials chain. It took a while for management to catch on. When they did, they used benchmarking to improve their customer billing operations. They succeeded, and customer satisfaction has never been higher.

- *Underestimating timing issues.* A benchmarking consultant colleague once "planned" a study for an electric utility that comprised roughly the following sequence of events.

Week 1: Identify three comparable plants for site visits.
Weeks 2 through 4: Visit (2 days each) comparable plants.
Week 5: Write report.
Week 6: Kick off implementation.

What's wrong with this picture? There are many things wrong, but the primary one is pretty simple. Managers of the target plants did not have any incentive to move quickly. "You want to come visit next week? For 2 days? Are you kidding? Don't you know we are trying to run a business here?" Needless to say, the benchmarking team was not in a position to exert any influence to speed things along at the target plants. The benchmarking site visits did take 2 days each, but the first was not performed until week 5 and the last two followed. *Lesson:* Plan for some cushion in timing-related issues, especially those dependent on people outside of your own organization.

The rest of this chapter is devoted to helping you plan your study well and avoid making these and other mistakes. The first issues addressed are the formation of the benchmarking team and documentation of planning in general. Then the first three steps in the eight-step benchmarking process are covered. Finally, a couple of points are discussed that should make your life easier—framing the report and identifying your prisoners.

Planning Your Study— Seven "To Do" Items

When you are planning a benchmarking study, seven items should be on your "to do" list:

- Determine which activities to benchmark.
- Identify the benchmarking team.
- Schedule the study.
- Determine the key factors to measure.
- Identify target organizations.
- Frame your report.
- Identify your prisoners.

The order shown is not necessarily the order in which you will perform these tasks, but all should be performed if you are going to have a well-planned study. For example, scheduling the study might be done after you have determined the key factors to measure. It might also be revised after you have identified the organizations you want to benchmark. Don't let the order restrict you. Just be sure to consider all seven items.

Determine Which Activities to Benchmark

Identifying areas for improvement is not difficult for most organizations. Prioritizing them may be. When all else fails, listen to your customers. Assuming that your organization has limited resources to devote to benchmarking and that being world-class in coffee room ergonomics is not on this year's "to do" list, the main criterion in selecting activities for benchmarking is to get the most for your money. If you only have resources to do three benchmarking studies per year, benchmark areas where improvements will help your organization's performance.

Activities that Increase Explicitly Your Organization's Value

For example, the greater percentage of your total costs that an activity or purchased input represents, the greater can be the impact of any savings generated by benchmarking that particular activity or input. If materials costs represent 60 percent of your total cost chain, a small percentage savings in materials cost will make significant savings to your bottom line. This idea is as old as management science, but it seems to be forgotten on a regular basis.

A ready-mix concrete producer's cost chain before benchmarking looked like the example in Fig. 6-2. Profits were down and decreasing further because competitors were winning competitive bids with prices below the company's cost to pro-

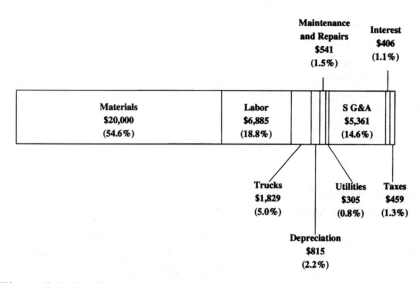

Figure 6-2. Ready-mix concrete cost chain.

duce. Management knew that costs had to be reduced and wanted some big savings fast. Analysis of the company's cost chain showed plainly that a small decrease in materials costs and direct labor would result in significant savings on a per-yard basis and make a large contribution to the bottom line. And preliminary analysis of the company's cost chain vis à vis the industry average cost chain, although rudimentary, showed that significant savings probably were possible in the selling, general, and administrative areas, too. Management benchmarked competitors' materials costs by collecting and analyzing data about sources and costs of purchased materials, quality of material used in specific mixes, and levels of cheaper chemicals used, when possible, in place of costly cement. Management found that their reputation for providing the highest-quality concrete in the market was being carried to a costly extreme by the company's engineering department, whose members were overdesigning almost every mix of concrete. In lay terms, they were frequently building sidewalks that were as strong as concrete high-rise buildings. Savings in the direct labor area were less promising, but management was pleased to discover that their direct labor force was relatively efficient.

Management also benchmarked competitors' selling, general, and administrative (S,G,&A) organizations and found that most were far more efficient with S,G,&A workforce than the company. Through attrition and realignment of administrative tasks, the company was able to reduce S,G,&A head count and, combined with new materials policies, was able to cut the per-yard cost of concrete production (previously $57.68) by $4.27, a significant savings in a short time.

Not all benchmarking studies are cost studies, however. Remember Porter's three generic strategies (Chap. 1)? Not all companies compete on cost. In fact, most don't. While it is very important to keep costs in line, most companies try to differentiate themselves somehow because they realize there can be only one lowest-cost competitor in an industry. The greater the value perceived by your customers in one of your organization's activities, the greater the impact of any performance improvements generated by benchmarking.

Management of an electronic design automation (EDA) software firm saw the impending maturation of the industry and realized that differentiating their software products from their competitors' was becoming increasingly difficult. The players remaining in the industry were very close technologically, and major technical breakthroughs—those that would result in long-term, sustainable differentiation—were happening with less frequency. Management decided to change the rules of the industry and differentiate the company by providing customers with world-class after-sales service and support. They benchmarked their major competitors' after-sales service and support processes to understand exactly the "floor" for being the best in the industry.[1] Their short-term goal was to be the best in the industry. And they benchmarked IBM, Xerox, and Hewlett-Packard to understand what world-class meant as it related to after-sales service and support. Customer satisfaction ratings in this area improved dramatically in the first 6 months after implementation of improvements. At this writing, plans are considered "on track" to meet customer satisfaction targets and to continue improvement of the process.

Activities Where the Climate for Change within Your Organization Is Right

Benchmarking is sometimes done for less tangible reasons, too. The hungrier an organization's people are for improvements, the more likely benchmarking-driven improvements will be successful. It may be difficult to quantify the value of this, but despite being philosophically at odds with the whole concept of benchmarking (meaning hard measurements), some organizations benchmark because they realize the climate for change is right and that benchmarking can help.

Morale was at an all-time low among service technicians at an automobile dealership chain in the New York area. A regional consumer's guide had surveyed car owners about the service

---

[1]And, like experienced benchmarkers who believe that competitors never stand still, aimed well above it.

functions of automobile dealers in the tristate area. The company's service was ranked in the bottom 10 percent. The technicians were, in the service manager's words, "mad as hell and not going to take it any more."[2] Management benchmarked the service functions of the five top-ranked dealerships and, with the delighted participation of the service technicians, developed and began implementation of a plan to move into the top quartile by the next survey.

Some companies determine the activities they plan to benchmark in a particular period based on the results of internal polling of employees. This approach does two things at once:

- It ensures that areas in need of improvement are considered.

- It is a good indicator that those who will be responsible for making any benchmarking-driven improvements will buy into the improvement program.

Remember, the ultimate goal of any for-profit business is to enhance shareholder value. Given limited resources to devote to benchmarking, benchmarking projects may be evaluated like capital budgeting alternatives, where those that appear able to add the most value are chosen. Some organizations, such as Xerox in the early 1980s, take a maniacal approach to benchmarking and benchmark practically everything that they do. For Xerox, a company in dire straits at that time, this across-the-board approach worked. For most organizations, especially those just beginning to use benchmarking, an approach like this would be doomed from the beginning. It is tantamount to trying to do everything better at once. Focusing on the key factors for success in the industry while deciding which activities to benchmark and keeping the three guidelines above in mind will provide the best opportunity for success in your benchmarking efforts.

Some organizations, such as Praegitzer Industries, Inc., a manufacturer of printed-circuit boards, conduct internal surveys to determine activities to benchmark.[3] Assuming you do not have unlimited time or money to benchmark everything on

---

[2]The service manager admitted that he was a movie buff.

[3]Whiting, R., "Benchmarking: Lessons from the Best-in-Class," *Electronic Business*, October 7, 1991.

Following is a list of potential areas for benchmarking in the next 12 months. Please allocate a total of 100 points to the choices. You may allocate as many or as few points to the choices (no negatives, please), as long as your total equals 100.

	Points
Billing	———
Customer Services	———
Engineering	———
Field Service	———
Finance	———
Marketing	———
Product development	———
Purchasing	———
Sales	———
Total	100

Figure 6-3. Sample forced-choice survey.

the list, one of the best techniques to use is often the forced-choice survey, where potential choices for benchmarking are required to weight each choice. Typically, decision makers are given 100 points and told to allocate them, as they see fit, to the possible choices. They can allocate all 100 points to a single choice, if they wish, or any other variation. Some basic mathematics skills are required to tally the inputs, but the result is a prioritized listing upon which the decision makers can usually reach consensus. Figure 6-3 is an example of a forced-choice survey used by managers at a telephone company.

Another framework used to analyze potential areas for benchmarking is the *competence gap matrix*. Activities at which the organization must excel to be competitive are mapped along two dimensions—the organization's level compared to its competitors' and the level required for competitive advantage. Figure 6-4 is a competence gap matrix for a software firm. Management used the matrix to communicate to the organization the urgency of making improvements in various key processes. The "actual" level of the company's competence compared to its competitors and the required level were deter-

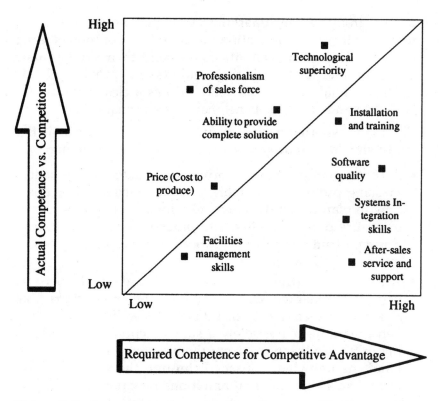

Figure 6-4. Competence gap matrix for software company.

mined by surveying customers, competitors' customers, managers, and employees. The lower right-hand corner is the "hot corner," i.e., it contains the areas where management should focus benchmarking efforts.

Identify the Benchmarking Team

Inclusion of the appropriate people on the benchmarking team will increase the likelihood of the ultimate success of your project. There are no inviolable rules in team formation, but experience shows the following guidelines to be generally helpful:

A core group of benchmarking gurus may be a good use of resources if your organization plans to do six or more studies per year. This group can efficiently build their skills in data collection and analysis if they undertake approximately six studies annually. They can serve also as a clearinghouse for both data and "how to" benchmarking information.

The team, whether a core group or ad hoc for a particular study, should include *at least one* of each of the following:

- A *benchmarking guru* who brings knowledge of the benchmarking process to the table. This knowledge will include how to plan the study, how and where to obtain data, how to ensure data reliability, and other issues that will arise during a study and that an experienced person can resolve more easily than a novice.

- A *line person* extremely knowledgeable about the activity being benchmarked. One of the strengths of benchmarking as a management tool is that it is practiced by line people—benchmarking is not solely a staff function. By participating in the study, people from the line not only absorb first-hand the knowledge learned through the study but also overcome skepticism that ambitious new performance levels can be achieved.

- A *change leader* who has whatever qualities or authority necessary in your organization to ensure that improvements are actually made based on the results of your benchmarking study.

These three roles can theoretically be played by a single person. In fact, the line person and change leader are frequently the same person.

The size of the team depends on the complexity of the activity being benchmarked, the number of companies to be studied, and the time frame within which the study should be completed. Teams of 3 to 6 people are very common and seem to work well. Large teams, say of 10 or more, seem to bog down, as one might expect in a democratic society. Generally, the team should have enough members to permit completion of the planning and data collection and analysis phases of the study within 10 to 12 weeks. This period of time seems to work well for a couple of reasons. First, when benchmarking, you are frequently at the mercy of others' schedules—people outside your organization. They do not have the same motivation to complete your study that you have, and you cannot exert influence on them as you might if they were insiders. Especially with site visits, it is unreasonable to believe you can complete a rigorous benchmarking study in just a few weeks. Second, there is the other side of the timing issue. By dragging a study out for much longer than 3 months (I have heard some war stories of studies that were planned to last as long as a year) you run the risk that

- Team members will not be around to complete it.
- People will lose interest in the study, and it will wither on the vine.
- Data collected early will be out of date by the end of the study.
- Team members will be very inefficient in execution due to a lack of focus.

This last item can be a real problem. Team members assigned to a study in addition to their normal responsibilities often find excuses to do anything but data collection. Data collection is difficult, tedious work, and data collection tyros will often find anything else to do, subconsciously of course, if it will keep them from making ten more phone calls seeking data and getting no response or, worse, rejection. Full-time responsibility and a 3-month deadline can go a long way toward obviating that problem.

It is often a good idea to have a cross-functional team composition. This ensures that improvements in one area are not wasted because of status quo in other complementary areas. For example, in the example cited earlier, the large regional chain of building materials stores benchmarked Nordstrom, Federal Express, and Home Depot, among others, to learn ways to improve customer service. The team consisted entirely of front-line types such as sales and delivery people. New standards were set in what turned out to be all *visible* aspects of customer service. Back-office billing problems were identified fortuitously and might have been missed under different circumstances. This area would have been easily identified if a cross-functional approach had been taken.

Schedule the Study

Once the benchmarking team has been established, like anything else, a well-thought-out and well-documented plan of action will ensure you have covered all the steps and designated responsibility to get them accomplished. (A Gantt chart that includes all steps in the benchmarking process and the discrete activities within each step is a valuable tool to use to ensure the ultimate success of your study.) The benchmarking team leader should take responsibility for holding all members of the team to completion of their tasks by the dates set. Letting dates slip by without completion of various tasks is usually detrimental to the project's entire likelihood of success. Figure 6-5 is a timetable for

Eight Benchmarking Steps	Step Activities	Respon-sibility	Total Budget (Hours)	Budget/Actual By Week													Actual Hours
				1	2	3	4	5	6	7	8	9	10	11	12	13	
Determine Which Activities to Benchmark	Poll Management																
	Poll Employees																
	Compile Results																
	Select Highest-Priority Issues																
	Determine Activities to Benchmark																
Determine Key Factors to Measure	Brainstorm Selected Activity																
	Define Performance Measure																
Identify Best-in-Class Companies	Initial Compilation of Potential Targets																
	Analysis of Potential Targets																
	Final Selection of Benchmark Targets																
Measure Our Performance	Gather Performance Data																
	Analyze Performance Data																
	Determine Data Needed from Target																
Measure Their Performance	Gather Performance Data (Initial)*																
	Analyze Performance Data (Initial)																
	Conduct Site Visits																
	Analyze Performance Data (Final)																
	Compare Performance Ours vs. Theirs																
	Analyze Performance Gap																
	Verify Gap Causes																
Develop a Plan to Close Gap																	

Figure 6-5. Timetable for benchmarking study. (*See Fig. 6-6 for further detailed plan of this step.)

Step Activities	Total Budget (Hours)	Week													Man-power required	Respon-sibility
		1	2	3	4	5	6	7	8	9	10	11	12	13		
Conduct Lit Search	20						◄►									
Analyze Lit Search Data	40					◄—►										
Interview Industry Analysts	32						◄►									
Interview Customers	40							◄—►								
Interview Suppliers	40							◄——►								
Interview Former Target Employees	24									◄——►						

Figure 6-6. Timetable for initial gathering of performance data.

completing a benchmarking study, and is general enough to be applied to many types of studies. Each box is split in half, so you can fill in your budget (top left half) and actual times. Tracking your time to this level of detail is a good idea if you plan to institutionalize the benchmarking process and perform many studies. You will be able to determine those areas in which your organization is getting more efficient over time and to plan future studies better. Tracking all your time may be overkill, however, if you only plan to perform occasional studies.

Of course, each step within a macro plan such as shown in Fig. 6-5 is composed of many smaller steps, all of which should be carefully planned, documented, and managed to ensure their ultimate completion. Figure 6-6 shows the detail for initial gathering of performance data for the same study.

Determine Key Factors to Measure

Determining which activities to benchmark in your organization is the first step and is usually straightforward if the above guidelines are followed. A. Steven Wallek of McKinsey & Company offers the following advice:

A few truly comparable numbers are worth much more than volumes of poorly understood, unverifiable data. Cost, quality and timeliness are the three variables to be measured; always measure all three.

Total dollars per unit or per ton is the place to start for manufacturing processes. Fully loaded costs per transaction or per time period can usually be derived for managerial processes. Unless the benchmarking team starts with total costs before it breaks them down by process or by activity, it may neglect some very important overhead and burden charges that are reported differently in different accounting systems.

Quality measures should capture the errors, defects and waste attributable to an entire process and express them relative to the total output achieved. Since defects tend to cascade down a chain of processes, becoming ever more expensive to correct, it is usually better to use failure rates—rather than best quality numbers—to measure quality performance.

Timeliness indexes should be as comprehensive as possible, starting with an initiating event—such as a customer placing an order—and following through to the moment a company is paid for results. Often the ratio of productive time to total time available is an eye-opening measure.[4]

Cost, quality, and timeliness are, in a different sequence, another way to say better, faster, and cheaper. The key is to focus on those measures that are going to tell you the most about the process you are studying. The Pareto principle—named for Vilfredo Pareto, the Italian economist who observed that most of the world's wealth is held by only a few people—really holds true with respect to this issue. Here, most of the insight is in a few key measures.

The following are some key guidelines to follow when you are determining the key factors to measure, i.e., metrics:

- *Keep focused.* It is easy to hear the siren song of benchmarking and begin measuring every input that could contribute to success in a particular function, process, or activity. Not only is this costly, but also it could lead to data overload, the benchmarker's nemesis. Too many data to analyze. Try to focus your

[4]A. Steven Walleck, "Manager's Journal," *The Wall Street Journal,* August 26, 1991, p. A8. Reprinted with permission.

measurement efforts on the smallest number of measures that will enable you to make the improvements you require.

- *Keep an open mind.* Imagine the learning that comes from going into a benchmarking study thinking you wanted to measure X and discovering that the best-in-class measure Y instead! A computer manufacturer wanted to measure board line yields at its competitors and discovered that the best shunned the yield measurement because it was too broad to be of any practical use. Instead the competitors measured each process within the line and were most concerned with defects in solder joints—the joints where chips and other components are soldered to the printed-circuit board—which they measured per million. So defects per million solder joints became the new measure used to gauge the board line manufacturing process.

- *Most importantly, remember that the underlying reasons for superior performance are usually more important than the performance numbers themselves.* Benchmarking is not just a numbers exercise. From the example immediately above, consider the following conversation, which actually took place.

> ENGINEER 1: Our computer board line yield is only about 40 percent. What did you learn about the yield of the best-in-class?

> BENCHMARK CONSULTANT: The best-in-class do not measure yield. They measure *defects per million,* and in this case defects per million solder joints, the most common cause of defective boards. Using that measure, some of your colleagues have calculated that your 40 percent yield translates to approximately 1200 to 1500 defects per million solder joints.
>
> The best-in-class, including Fuji-Xerox, Hewlett-Packard, and Motorola, routinely achieve fewer than 20 defects per million solder joints and quite frequently achieve rates of 2 to 5 defects per million solder joints.

> ENGINEER 1: Well, what is it? 2? 5? 20? Some other number?

> BENCHMARK CONSULTANT (trying to determine a tactful way to make the point to the client): Hmmm...

> ENGINEER 2 (interrupting, and not at all worried about tact): Who cares? If we're at 1500, we could cut our defects in half,

then cut them in half again, then cut them in half again, and we still wouldn't be close. What we need to know is, How do they do it?

Exactly.

Identify Target Organizations

The possibilities seem endless. Your strongest competitor? All your competitors? Latent competitors? Or perhaps a company from the seemingly hundreds of U.S. companies that have been labeled best-in-class or world-class at some activity? Did we consider what are likely thousands of other companies in the rest of the world that have been labeled best-in-class in the languages that their managers speak?

There are generally four groups of companies that you might consider in selecting benchmarking candidates:

- Current direct industry competitors

- Latent competitors, including those in your industry but not currently in your market (e.g., Japanese automobile manufacturers in the U.S. bus market) or those not in your industry at all who could enter (look at what AT&T Universal Card Services is doing in the credit card market)

- Best-in-class groups from within your own organization

- Best-in-class companies from other industries

Which organizations you study depends in large part on what you are trying to improve. If you believe your production costs are too high, you will likely study your direct competitors' in a cost benchmarking study to determine their costs, and perhaps some latent competitors if comparisons can be made on an apples-to-apples basis. (Cost benchmarking is covered later in this chapter.) But if you want to improve in an area where none of your competitors is particularly strong in an absolute sense and you want to leapfrog them all with some new practices, it is likely that your learning will come from outside your industry, from companies that are world-class in par-

Direct Competitors

- United
- American
- Delta
- Alaska

Studied to establish absolute minimum service standards to be competitive in region.

"Latent" Competitors

- Scandinavian
- Singapore
- JAL
- ANA

Studied to understand best airline practices, despite the fact that these airlines did not currently fly the regional airline's routes.

Other Best-in-Class

- AT&T Universal Card Services
- Nordstrom

Studied to understand world-class customer service standards by acknowledged leaders.

Figure 6-7. Potential companies to be studied by a regional U.S. airline. Objective: Improve key facets of customer service.

ticular activities that are also performed in their industries. Figure 6-7 is the list developed by an airline seeking to improve its customer service. Note that the list of potential benchmarking targets, in this particular instance, included companies from three of the above categories.

Following are some general guidelines that will help you select the right companies to benchmark.

Direct and Latent Competitors

- *Always consider your direct competitors.* Foolish is the manager who chooses not to study direct competitors because they are not among what he or she considers best-in-class benchmarking candidates. If you are benchmarking firms outside your industry, your competitors may be doing so as well. If performing a rigorous, full-blown benchmarking study on them seems like overkill, you should at least perform competitor analysis of the prebenchmarking variety (see Chap. 1 for further discussion of this topic). Management of an electronic design automation software firm that benchmarked IBM's, Xerox's and Hewlett-Packard's after-sales service and support also benchmarked their two main competitors. They learned that one competitor was also beefing up its after-sales service and support capabilities in an effort to change the rules of competition in the industry. This enabled management to aim their short-term performance goals higher than they might have otherwise—and emerge victorious in the competition.

- *Ask your customers.* Ask them who in your industry is best at the activity you wish to improve. Customers will know. But better than asking your customers may be asking people or companies that *don't* buy from you. This type of inquiry has to be made discreetly to avoid tipping off the competition to the fact that you are on the improvement trail, but you probably can learn a lot from your competitors' customers.

- *Ask your employees.* Ask your employees, especially your sales people, delivery people, and service representatives. They are in the field, going head to head with the competition every day. They usually have pretty good ideas about who is best and who is not in a wide variety of areas.

Benchmarking within Your Own Organization

Benchmarking within your own organization, also known as *internal benchmarking,* is often done by large organizations with widespread operations. These organizations, which may have many thousands of employees spread throughout the world, frequently contain pockets of excellence in various processes that are hidden away in remote locations just waiting to be discovered. For example, a Big Six accounting firm[5] uses biweekly

---

[5]The Big Six accounting firms are Arthur Andersen, Coopers and Lybrand, Deloitte and Touche, Ernst and Young, Peat Marwick, and Price Waterhouse. The Big Six used to be the Big Eight, but they do not do much benchmarking and have suffered from a deteriorating industry structure. Thanks to lawsuits, increasingly sophisticated and price-sensitive clients, and an endemic inability to add much value through their auditing practices, the Big Six may continue to be subject to consolidation.

time reports to accumulate billable hours and out-of-pocket expenses generated by its partners and employees. Unless these reports are submitted on time and correctly by the partners and employees, proper billing of clients cannot take place, unbilled receivables mount, and the firm incurs greater capital charges and likelihood of uncollectible accounts than it would if the process ran smoothly. One would think that getting meticulous CPAs to complete and submit accurate time reports when due would be easy, but it's not. So the firm benchmarked its field offices nationwide and was astounded to learn of some very workable practices that a few offices used to promote timely and accurate submission of reports.

Internal benchmarking is typically just the first phase in a larger, outward-looking benchmarking study, but can uncover some valuable process improvements with a minimum of resources spent and little or no difficulty in obtaining cooperation from the best-in-class. Quite frequently, people in a large organization who have followed certain practices unnoticed for a long time have been trying unsuccessfully to share their knowledge and are only too pleased to do so when finally asked as part of a formal internal benchmarking effort.

Other Best-in-Class Companies

- *Select companies with obvious strengths in the activity you want to benchmark.* This is getting easier to do as the practice of benchmarking expands. As more and more companies catch on to the power of benchmarking, the identity of best-in-class companies becomes more widely known. Pity the person in charge of distribution and warehousing at L.L. Bean, who has become, ex officio, a public figure. This widespread knowledge of best-practice companies has its downside too. As requests for information grow, sometimes exponentially, the likelihood of getting in-depth knowledge from them diminishes. They have businesses to run, too.

- *Look at Malcolm Baldrige National Quality Award winners.* Baldrige Award winners are, by and large, quality zealots who are usually extraordinarily gracious in sharing information. We have found our requests for cooperation from Baldrige winners to be met with great alacrity and professionalism, although the winners seem to be getting more sensitive about how they use their time.

- *Look for companies with comparable products, services, or processes in some important characteristics.* When a Big Six accounting firm wanted to improve its purchasing function, managers identified several firms that purchased large volumes of comparable items (travel, office supplies, office furniture, etc.) and benchmarked them. When a heavy equipment manufacturer wanted to improve its line changeover process, managers identified several other manufacturers of products with comparable physical characteristics (heavy castings, sheet metal, etc.) and benchmarked them. Learning how line changeovers were done at a high-technology manufacturer might have provided some valuable insight, but may have lacked comparability in important respects.

- *Look at industry publications.* Many industry publications include annual lists of the top companies in a particular industry or function. *VAR Business, Information Age, Industry Week, etc.,* can be good sources for identifying potential benchmarking targets.

- *Look at App. A.* Appendix A is a short compilation of leaders in different functional areas, processes, and activities that we have heard about in our travels. It is not meant to be definitive or comprehensive, but it might provide a good start.

Remember, when you are benchmarking companies outside the industry, the possibility exists for significant gains over current practices in a given activity. Do not get discouraged if you cannot secure cooperation from the absolute world leader in a particular activity (unless you have reason to believe your main competitor can). If learning from a nonindustry company, whether best-in-class or not, is going to allow you to dramatically improve performance and perhaps leapfrog past competitors, you have nothing to lose by doing it. Xerox learned a tremendous amount about warehousing and distribution by benchmarking L.L. Bean in this area.[6] Could they have learned enough about this area from Lands' End? Probably. As a matter of fact, has anybody ever compared the Lands' End and L.L. Bean warehousing and distribution functions? Could it be that the people at Lands' End really have warehousing and distribution nailed, too, and thank the Lord every day that they were not discovered by Xerox and hence can go about their business each day uninterrupted by requests to share data?... Have you ever tried to get L.L. Bean to participate in a benchmarking study? It is not done easily anymore, which is understandable. After all, the people at L.L. Bean are trying to run a business.

Remember, do not limit your benchmarking to direct competitors. What you can learn from competitors in many areas is less than what you can learn from companies outside the industry. And do not limit your search to companies in your immediate geographic area. Do you want to be best-in-class or best-in-Connecticut?

Also, when you are benchmarking companies other than your competitors, it is usually a good idea to identify initially more target companies than you think you will eventually study. A weeding-out process is often done, during which a number of potential benchmarking targets are reviewed preliminarily to identify those that appear to be

- Most comparable to the team's company in terms of the process to be studied.

- Most different from the team's company in terms of the process to be studied, as long as the end product or service is the same. This may open your eyes to a totally new way of doing things.

- Best of the initial candidates following some preliminary data collection and analysis.

---

[6]See Robert C. Camp, *Benchmarking—The Search for Industry Best Practices That Lead to Superior Performance*, Quality Press, Milwaukee, 1989, for a full rendition of the L.L. Bean story.

- Most willing to cooperate in the study, which is becoming a real issue as benchmarking becomes more popular. If you really want to benchmark a particular company that you suspect is a popular target for benchmarking, you may find that managers there just cannot give you the time you want. Most of the Baldrige winners are swamped with requests and simply do not have the time to provide one- or two-day site visits to everyone who requests them. You can help your chances in two ways. First, make it crystal-clear from the outset that you are doing real, rigorous benchmarking and not wasting everyone's time with industrial tourism. Second, come bearing gifts. Have something to offer the people you want to learn from, in the form of some special expertise of your own. Quid pro quo. I cover this more in the section on identifying your prisoners.

A few final words on target selection are warranted. Your first real data collection and analysis efforts begin during the selection process. You cannot honestly identify benchmarking targets until you have satisfied yourself that your targets are the "right" organizations, based on whichever combination of the above criteria makes sense for your study. The most important criterion in selecting targets is to choose those from which you will be able to learn, learn, learn.

Also, try to avoid encumbering the initial target identification process with too many rules. Many benchmarking teams begin with a large list of potential targets and narrow it by a significant factor before ultimately deciding whom to study. One of the Baldrige winners, in a study of leading practices in the use of customer satisfaction data, began with a list of almost two dozen potential targets and narrowed it to a field of six, which ultimately participated in the study. The large list included names of potential targets that the team developed by brainstorming supplemented with potential targets that came through such conventional media as E-mail from nonteam colleagues who knew about the study and a note from the CEO written on a letter he had received from the customer service center at a company he does business with. The large list was later narrowed, of course, but the point here is to be creative and open-minded when you are identifying firms to study.

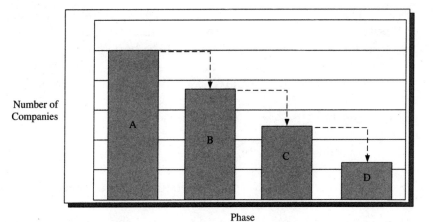

Figure 6-8. Target company attrition. Phase A—brainstorming list: The total population of potential target companies. Cost per potential target at this point is nil. Phase B—approach list: These are the companies the benchmarking team approaches about participating in the study. Some attrition or addition during this phase typically occurs, for the following reasons: Initial research indicates that brainstorming "gut feel" was inaccurate; initial contact to obtain information indicates aversion to participate by potential participant; and initial research identifies more good potential participants. Cost per potential participant at this point is low, so value of forced attrition is also. Rule of thumb: If a company does not appear to be a poor candidate, do not eliminate it from the list. Phase C—questionnaire-consortium participants: The population of companies that agree to participate in the study. Costs here include your time in securing their participation and expenses related to publishing, mailing, etc. Phase D—actual questionnaire/consortium participants: This number will likely be smaller than that of phase C. Attrition primarily due to unwillingness to participate by potential participants.

The chart shown in Fig. 6-8 is a typical progression of how the population of potential target companies becomes the population of actual targets participating in a study. Always plan for some attrition when you select targets for study.

Frame Your Report

Framing your report means writing a detailed outline after you have identified the targets but before you have begun in-depth external data collection and analysis. You should view framing the report as an important step in the planning process. At this point you have determined the activities to benchmark and the key factors to measure, and you have selected target companies. Framing the report is a valuable exercise that can be done on almost all benchmarking studies. In essence, it entails writing the report without the data. This exercise is valuable because it forces you to think through *in great detail* all the data you will need to complete the study, how the data will flow together, and where you have to go to find the data.

Framing the report may seem inefficient because the report will change—sometimes organically—before you have finished it. It is. Some unexpected discoveries during the data collection and analysis phase may make part of your data collection moot or may send you down an entirely new path. But the benefits from having thought through the study in great detail almost always outweigh inefficiencies.

One word of caution, however. Do not draw any conclusions when you are framing the report. Framing the report is a planning exercise, but planning conclusions before you have even begun data collection and analysis is neither honest nor wise.

Identify Your Prisoners

An *exchange of prisoners*[7] means that you have something of value to share with those you hope will cooperate or collaborate in a study you are performing. Identifying "prisoners" is reality for those who wish to perform cooperative benchmarking in the 1990s. Many of the better-known best-in-class companies have been inundated with requests to share by benchmarking teams interested in learning about their best practices. The volume of requests has forced some to become more discriminating in granting true benchmarking privileges, especially to those who either appear to be "industrial tourists" or have no prisoners to exchange. Having prisoners to exchange may simply mean having one or more best practices of your own—practices that oth-

---

[7]Many thanks to Mike Macfarlane, Vice President of Corporate Quality at Sybase, for this illustrative term.

ers can learn from. For example, one of the large credit card issuers offered data about its billing and collections process as an inducement to other organizations to get them to share data about customer service, telephony, and innovation.

There are three components to identifying your prisoners:

- Identify those areas in which your organization excels.
- Determine which you are willing to share with others and at what level of detail.
- Determine with whom you are willing to share them.

Do not labor under the misconception that every organization that has a best practice is willing to share it with the rest of the world. Even the Baldrige winners are not *obliged* to cooperate in benchmarking studies. Flattery is working with less frequency. But if you bring something to the table, you will enhance your chances of learning what you want to learn.

Of course, not all organizations have something to bring to the table. Lacking any truly superior practices to offer up in trade to potential benchmarking partners, some organizations bring the fact that they are organizing and running a collaborative study. Organizing and running a study is a great deal of work, and many organizations realize that doing this work and delivering value to all the organizations that participate in the study is often enough to get some world-class organizations to play. For example, one of the Regional Bell Operating Companies (RBOCs) recently organized and ran a study of leading practices in the use of third-party distributors and agents to sell a company's products or services. The use of independent sales agents by the RBOCs is relatively new, and managers at this RBOC felt they probably had more to learn from leading companies than offer them in this area. Nevertheless, they were able to enlist several leading computer and insurance companies to participate in the study because it was an area of great interest to them, too, and none of them had to be bothered with the work that went into organizing and running the study.

The seven "to do" items just covered apply on almost any benchmarking study. If you address each one every time you benchmark, your studies will go much more smoothly than if you overlook one or more items. Do not treat the list as all-inclusive, however. There may be other issues in a particular study that warrant your attention during planning.

The remainder of this chapter discusses some special considerations to keep in mind in planning a cost benchmarking study and a study of a process that is rather broad in nature.

A Note about Planning a Competitive Cost Benchmarking Study

A thorough cost benchmarking study of your competitors will not only enable you to find ways to become more cost competitive under your present system, but also should enable you to identify strategic alternatives in the form of different business systems, i.e., different ways of delivering value to customers in the industry. Many will agree that benchmarking competitors' costs is the most challenging application of benchmarking you can undertake. For starters, your competitors do not want you to do such a study. Obtaining direct information is usually difficult at best and seemingly impossible at worst; competitors seldom mirror precisely your own company in the use of technology and markets served (which makes meaningful comparisons more difficult to derive); and reconstructing cost chains from various cost drivers means combining a good amount of judgment and best guesstimates with incomplete and often inconsistent data from many sources. Multiply all these challenges by a large number of competitors and potential competitors, and the task can seem insurmountable. A well-planned methodology will remove many obstacles.

A good way to start a cost benchmarking study is with an analysis of the different ways that value can be delivered to customers in the industry. This analysis is not actually benchmarking but is a precursor stage that will help you

- Highlight and analyze the different ways companies compete in your industry.
- Identify different ways to approach the issue of satisfying your customers.
- Focus on the high-leverage-value activities in your value chain.

■ Identify and narrow your study to those competitors that will give you the greatest return through benchmarking.

The first step in this analysis is to construct the organization's value chain and analyze the costs of each link in the chain, typically on a per-unit basis (this permits you to make meaningful comparisons with the value chains of competitors later). Figure 6-9 shows a simplified value chain and aggregated costs for a ready-mix concrete producer.[8]

Next, identify the various options that exist or could exist for each link in the chain. Different companies deliver value in dif-

[8]Many thanks to Gene Ceccotti, CEO of Shamrock Materials, and his people for help with this section. Ready-mix concrete is a simple industry to understand, which is why it is used for illustrative purposes here and in other places in this book. However, this analysis can be applied to practically any industry, and should be, before launching into a costly and time-consuming cost benchmarking study of competitors.

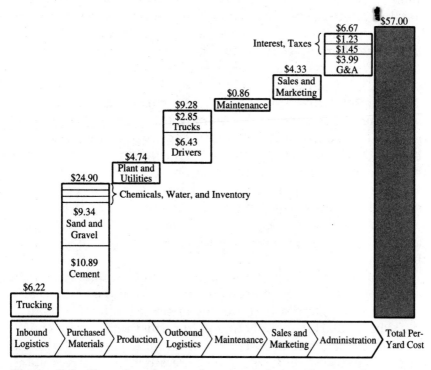

Figure 6-9. The value chain and aggregated costs for a ready-mix concrete producer.

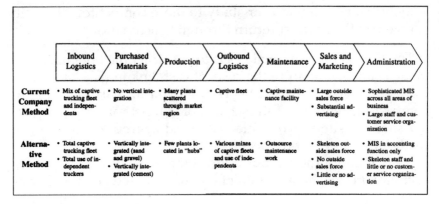

Figure 6-10. Alternatives in the value chain.

ferent ways. It's important to understand how each link in the chain *could* be delivered in your industry and how changes in one link affect the performance or cost of other links. Figure 6-10 continues the ready-mix concrete company example by identifying the alternatives that could be used.

Once you have identified the alternatives to delivering value in your industry, estimate how the various alternatives affect the total cost of a value chain comprising different permutations. In this exercise, compare the alternatives to your own value chain, and estimate the impact on total cost through various changes in the chain. Be certain to consider the impact of changes in one area on other areas in the firm. For example, using independent truckers for 100 percent of inbound logistics work might increase the cost of inbound logistics by 10 to 15 percent but *decrease* the cost of maintenance and interest expense. Figure 6-11 illustrates this piece of the analysis.

Remember, this step in the analysis should reflect the best judgment of managers of each area included in the value chain. You are not seeking a high degree of precision at this point—actual benchmarking will help you to further refine your understanding of the various costs later. Right now you want reliable estimates to use in understanding generally how different value chain configurations accumulate their costs.

	Inbound Logistics	Purchased Materials	Production	Outbound Logistics	Maintenance	Sales and Marketing	Administration
Current Company Method	$6.22	$24.90	$4.74	$9.28	$0.86	$4.33	$6.67
Potential Cost Differential, this link	+15 - 10%	-10 -15%	-5 - 10%	+5 - 10%	+10 - 15%	-15 - 25%	-10 - 15%
Potential Net of Costs of Other links	$.50	+$3 - 5	-$2 - 4	Insignificant	Insignificant	Insignificant	+$2 - 4
Importance of link	minor	major	major	minor	minor	major	minor

Figure 6-11. Estimated cost impact of alternatives in the value chain.

In this particular instance, the company's value chain comprised, with the exception of inbound logistics, the most expensive alternatives to configuring the entire chain. The company had made the decisions consciously over the years to provide superior quality and customer service, which resulted *unconsciously* in a buildup of costly approaches to executing every link in the chain. In most other industries, an assessment of the impact on quality and customer service would have to be made concomitant to the analysis of Fig. 6-11. In this case, however, the company had abundant evidence to support management's belief, developed over the years, that most customers would not pay a significant premium for what they considered a commodity product.

The final step in the prebenchmarking analysis, therefore, was to determine which competitors had value chains that included the alternatives considered to have major cost impacts, based on the preceding analysis. As it turned out, none of the competitors was vertically integrated. Two operated, however, with "skeleton" sales forces and little or no advertising and based their operations from a small number of hub plants. The company selected these two competitors for cost benchmarking as well as two others, whose value chains were configured substantially like the company's. These latter consistently beat the company in competitive bidding situations and were presumed to have lower costs. The first two were studied to learn with greater accuracy the advantages of differently configured competitors while the other two were studied to learn how the company might better its cost position within its existing value chain configuration.

Having developed a sophisticated understanding of the issues and having narrowed the field of potential competitors to benchmark, the company was now ready to perform the actual benchmarking. All four companies' present costs were studied. Projections were made of future performance levels, assuming hypothetical changes within their existing value chain configurations and hypothetical changes of the configurations themselves. For example, the company performed sensitivity analyses to determine the effects on each selected competitor's cost structure of a vertical integration move by either a cement or a sand and gravel producer. Projection of future performance levels is a requirement of benchmarking, especially cost benchmarking studies.

A Note about Planning a Broad Benchmarking Study

The essence of the benchmarking process—learning from others—can be applied in many ways. Benchmarkers are getting more ambitious in defining the scope of benchmarking studies, which usually adds to the complexity of the process being studied. Benchmarking a broad process, as opposed to just a piece of the process, can be done, but the process itself is frequently easier to understand and study if it is broken down into its various components. The disaggregation of the value chain of the ready-mix concrete company above illustrates the breaking down of a broad process into its components to make the study of the process more understandable and manageable. To illustrate this point by using another example, consider the following.

Managers at a leading financial services company wanted to use the benchmarking process to "improve their training function." The training capabilities at this organization were not deficient. In fact, their training department was, in many respects, among the best in the United States. Employees averaged over 2 weeks of training per year, training operations were rather efficient, and the company had received high grades from Baldrige examiners related to training and human resource development.

In a spirit of continual improvement, however, the managers set out to "benchmark training" with some of the leading training organizations found at U.S. companies. Benchmarking the entire training process is a much broader endeavor than just benchmarking a piece of it, say, benchmarking the techniques used by the leading companies in the area of developing computer-based training (CBT) programs. Benchmarking CBT would be easy—a single area, a single issue. Identify the best firms in the area of CBT development and study how they do it. But what would they learn if they benchmarked the entire training process? They did not know for sure, but they did know that there was a great deal of accumulated learning and expertise inside their target companies and that access to their counterparts at these other companies would likely provide some valuable improvements to their own training operations.

Because the entire training process is so broad, however, to facilitate its study, the benchmarking team decided to break it down into various bite-size pieces, i.e., the components of training. They ended up with the chain shown in Fig. 6-12. As you can see, the training process can be broken down logically into each of the components shown, which makes studying it much more manageable. Within each area a specific set of questions can be developed and addressed with target organizations.

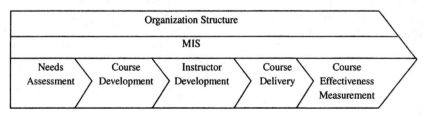

Figure 6-12. The training business system.

- *Organizational structure* encompasses all the areas of the training process. Information relevant to organizational structure covers where in the overall structure of the organization the training is situated, the number and experience of training professionals and support staff, the number of people trained, annual training goals/requirements, training budget, etc.

- *MIS*, like organizational structure, spans all the areas of the training process. MIS includes information relevant to measuring and reporting on financial, efficiency, and related measures in each of the primary training activities.

- *Needs assessment* addresses the organization's methodology for determining the need for training for all employees in areas relevant to their positions.

- *Course development* addresses the organization's methodology for developing and updating training materials.

- *Instructor development* addresses the organization's methodology for identifying, developing, and evaluating members of the training staff, including instructors.

- *Course delivery* addresses the organization's techniques and logistics for delivering training.

- *Course effectiveness measurement* covers the organization's methodology for evaluating training courses delivered and their impact on employees' performance in the workplace.

You should attempt such a disaggregation of any broad process that you decide to study. Note that the activities are linked in the sequence that they normally occur. What training do we need? How do we develop the course? How do we train the instructors? And so on. Any process can be disaggregated like this. The simpler the better. A chronological flow of events is typically the easiest and most understandable way to disaggregate the broad process. Design it; build it; sell it; service it—measure how you did.

7

Executing
Your Study

Data Collection
and Analysis

God is in the details.

Introduction to Data
Collection and Analysis

Many of today's leading business schools use the Socratic method, taught through case studies, to prepare students for the real world. Cases usually present a plethora of data in the form of text, graphs, tables, and other exhibits, and students are expected to sift through it all in 2 to 3 hours and converse intelligently and confidently about their particular "solutions" during class. Some overwhelmingly salient and compelling piece of data is often buried in an exhibit or footnote somewhere in the middle of the case and, with a little precision and luck, an aspiring MBA student will locate it, develop it, and score big with it during the class discussion. Case studies are great for teaching students to analyze data and think about managerial issues. It is too bad that the real world doesn't present itself wrapped neatly inside a 24-page case study.

Data acquisition and data analysis in the real world are hard work. Data never come packaged neatly, and different sources often provide widely disparate data about the same topic. You need a very healthy tolerance for ambiguity and a strong sense of perseverance in collecting and analyzing data. Data will be less than perfect and will age quickly, so you need always be willing to estimate and make decisions with the best available data at any point. The alternatives are commonly known as data addiction or analysis paralysis.

Some data are inherently easier to collect than others. Our experience is similar to that of most other veteran benchmarkers in a broad sense, and you may want to keep the following two general guidelines in mind when planning your data collection:

- It is usually easier to collect data from noncompetitors than from direct competitors.

- It is usually easier to collect data about noncore activities than about core activities. So, in many industries, it may be easier to collect benchmark data about human resources issues or MIS issues than about manufacturing or production costs.

This principle is displayed graphically in Fig. 7-1. Although it is easy to understand, many people forget to consider it when planning the timing of their studies.

Finally, a benchmarking study is often characterized as a jigsaw puzzle because it is complete only after many, many pieces have been sorted and arranged and a clear picture has emerged. And let's face it, when studying competitors, you might learn a seemingly useless fact from one interview and another seemingly useless fact from another. But taken together, these data turn into *information*. And *that's* what benchmarking is all about.

For example, in a recent study, the benchmarking team of a computer manufacturer learned the following pieces of information from separate sources at a competitor:

Source 1:   "We have 257 support personnel in our sales function in the western region."

Source 2:   "Our direct sales-to-support ratio is about 4 to 1."

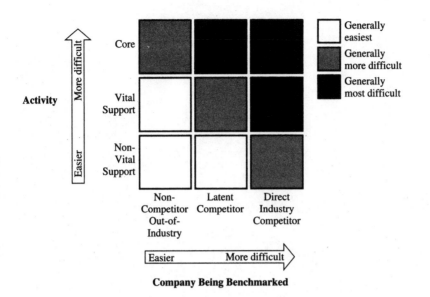

Figure 7-1. Activity versus ease of data collection.

Neither source would disclose how many direct salespeople worked in the western region. Taking these two facts together, however, the benchmarking team was able to piece together the jigsaw puzzle. Relentless pursuit is what's required in data collection. But you *can* do it.

Enough pep talk. This chapter will address the middle two steps in the benchmarking process, as shown in Fig. 7-2. Let's look at where data may be obtained and some useful items to consider when analyzing them.

Data Collection

First and Last—Internal Data Collection

Because one of the primary objectives of benchmarking is to compare your performance to that of competitors or other best-in-class companies, at some point during the study you will

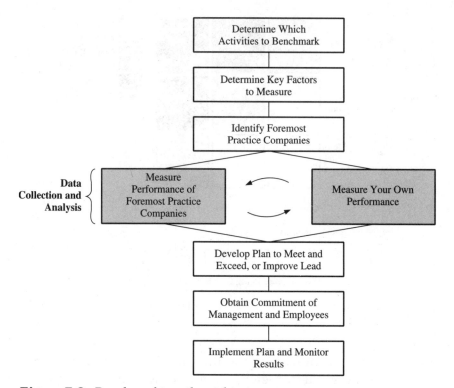

Figure 7-2. Benchmarking, the eight-step process.

have to collect data about your own company. This is typically done prior to collection of target companies' data. Usually your data are more accessible than data from other companies.[1] This forces the benchmarking team members to work through their own process at a level of detail they will need when collecting outside data. Study of your company's data confirms exactly what outside data are needed. Figure 7-3 is the internal data collection framework for a cost benchmarking study of the cost of goods sold for a basic manufacturing operation. The

---

[1]But not always. Managers who conducted a benchmarking study at an electric utility told of spending over *6 months* trying to get their corporate accounting department to correctly analyze operations and maintenance expenses per megawatthour for one of their plants. This egregious example of corporate bureaucracy and lack of intrafirm cooperation is an extreme example of an unfortunately common problem. Make sure you have considered internal bottlenecks when planning your study.

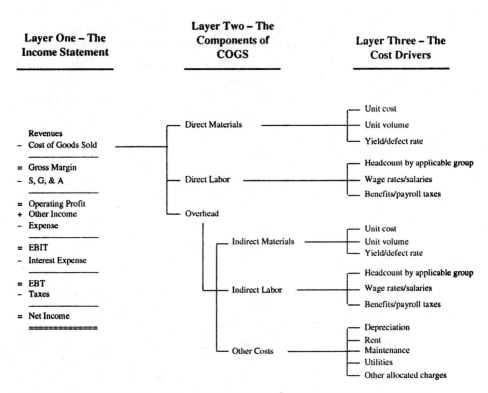

Figure 7-3. "Peeling back the layers" to go from income statement to underlying cost drivers, for Cost of Goods Sold.

sequence of "peeling back the layers of the onion" to go from the income statement to underlying drivers that collectively comprise the income statement is a necessary step in most cost benchmarking studies.

The process of collecting your own internal data may need to be repeated after external data are collected. The team may find that target companies measure their processes differently. It is usually easier to go back and measure your own process in the same manner as target companies measure theirs than to get target companies to measure their own process using your measurements. In Chap. 6, there was a story about computer hardware manufacturing managers who wanted to know optimum yield levels for their board line manufacturing process. Their own internal data collection, using their own measurement, indicated that they were at approximately 40 percent average yield in the process. Their external data collection efforts revealed that the best-in-class companies measured defects per million solder joints, not yield, which meant the company had to go back to measure their board line manufacturing process again, using this new measure. Although time-consuming, this remeasurement was part of the overall lesson learned about board line manufacturing at world-class companies. Defects per million is used by the world-class companies because it provides their managers with better information than yield does.

External Data Collection: Published and Electronic Data Sources

Data come from a variety of sources, from published sources to personal interviews and points in between. Appendix B is a listing, by industry, of some sources of information for the major industry classifications in the United States. Some of the more common published data sources are included in Fig. 7-4.

- Annual reports
- 10-K reports
- Prospectuses
- Dun and Bradstreet reports
- Trade journals (e.g., *Telephony, Computerworld, Electronic Business*)
- General business periodicals (e.g., *The Economist, Fortune, Business Week, The Wall Street Journal*)
- Company newsletters
- Wall Street analysts' reports
- Competitors' sales literature
- Competitors' local newspapers
- Government/regulatory filings
- Union agreements
- Industry directories
- Industry studies (e.g., Frost & Sullivan)

Figure 7-4. Common published data sources.

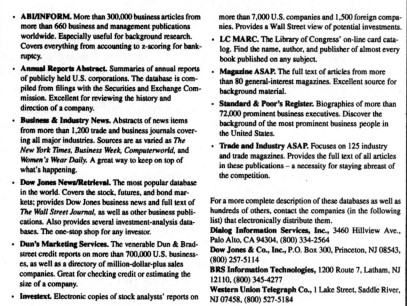

Common "Live" Data Sources

- **ABI/INFORM.** More than 300,000 business articles from more than 660 business and management publications worldwide. Especially useful for background research. Covers everything from accounting to z-scoring for bankruptcy.

- **Annual Reports Abstract.** Summaries of annual reports of publicly held U.S. corporations. The database is compiled from filings with the Securities and Exchange Commission. Excellent for reviewing the history and direction of a company.

- **Business & Industry News.** Abstracts of news items from more than 1,200 trade and business journals covering all major industries. Sources are as varied as *The New York Times, Business Week, Computerworld,* and *Women's Wear Daily.* A great way to keep on top of what's happening.

- **Dow Jones News/Retrieval.** The most popular database in the world. Covers the stock, futures, and bond markets; provides Dow Jones business news and full text of *The Wall Street Journal,* as well as other business publications. Also provides several investment-analysis databases. The one-stop shop for any investor.

- **Dun's Marketing Services.** The venerable Dun & Bradstreet credit reports on more than 700,000 U.S. businesses, as well as a directory of million-dollar-plus sales companies. Great for checking credit or estimating the size of a company.

- **Investext.** Electronic copies of stock analysts' reports on

more than 7,000 U.S. companies and 1,500 foreign companies. Provides a Wall Street view of potential investments.

- **LC MARC.** The Library of Congress' on-line card catalog. Find the name, author, and publisher of almost every book published on any subject.

- **Magazine ASAP.** The full text of articles from more than 80 general-interest magazines. Excellent source for background material.

- **Standard & Poor's Register.** Biographies of more than 72,000 prominent business executives. Discover the background of the most prominent business people in the United States.

- **Trade and Industry ASAP.** Focuses on 125 industry and trade magazines. Provides the full text of all articles in these publications – a necessity for staying abreast of the competition.

For a more complete description of these databases as well as hundreds of others, contact the companies (in the following list) that electronically distribute them.

Dialog Information Services, Inc., 3460 Hillview Ave., Palo Alto, CA 94304, (800) 334-2564

Dow Jones & Co., Inc., P.O. Box 300, Princeton, NJ 08543, (800) 257-5114

BRS Information Technologies, 1200 Route 7, Latham, NJ 12110, (800) 345-4277

Western Union Telegraph Co., 1 Lake Street, Saddle River, NJ 07458, (800) 527-5184

Figure 7-5. Common on-line data sources. (Source: Ken Landis, "A Decision Maker's Guide to Business Databases," *Macuser,* August 1989, p. 93.)

Some of the more common electronic data sources are described briefly in Fig. 7-5. Many of the articles, etc., from the published sources listed in Fig. 7-4 can be accessed electronically.

The process of getting data from published sources is usually referred to as the *literature search,* or "lit" search for those in the trade. A quick look at the lists above reveals the obvious—not too many detailed competitive secrets are going to be found in these sources. That's true. And while they typically provide only 20 to 30 percent of the data needed to perform subsequent analysis, your project is doomed without these data. If you do not build your foundation of knowledge about target companies, especially those with which you are not very familiar, subsequent conversations with important and valuable live sources will just be "fishing trips."

Published sources *must* be accessed before the real learning in most benchmarking studies can be done. Sometimes you hit a gold mine. Prospectuses are a wealth of information on a relatively macro level. You should always review any current ones, if available, for a company you are studying. A software firm that was contemplating entering the systems integration/information technology industry was conducting a benchmarking study of three top firms in that industry and a smaller software consulting firm (revenues approximately $100 million), to develop a sophisticated understanding of the four firms' strategies and, through induction, the keys to success in the industry. Lots of published data in various trade journals were to be found about the three big players, but not many articles had been written about the smaller firm. However, the smaller firm had recently held a public offering of common stock, and its prospectus was loaded with very detailed descriptions of industry success factors and the company's operating plans to compete. A gold mine, indeed.

It is important to stay focused during the collection of published data. It is only natural to want to learn everything possible about the company or companies you are studying, but that may be extremely time-consuming—and of little extra benefit. Avoid reading *A History of the Petroleum Industry* if all you really want to find out is last year's raw materials cost of one of the oil companies' chemical subsidiaries.

A Note about Who Should Conduct Literature Searches

There are professional services with access to on-line databases and published periodicals directories, and they will conduct a literature search for you for a nominal fee. Using topics, companies, and key words, they compile a list of abstracts of articles. You review the abstracts and identify the articles you want, and they supply them. The people who work at these services are often trained librarians, and they can conduct a literature search very efficiently. That's the good news.

The bad news is that they are not usually privy to the objectives of the benchmarking team and, even if they were, cannot usually exercise the judgment in searching through published literature that a member of the benchmarking team could. A benchmarking team, developing a sophisticated understanding of the primary competitor's quality characteristics, strengths, and weaknesses, asked for a literature search of titles that included key words such as *quality* and *customer service*. They got many articles about the competitor that included the key words in the titles. Other important articles were totally overlooked—until brought to their attention by the executive vice president to whom the benchmarking team was reporting. One item that was overlooked was a feature article in an industry journal entitled *[XYZ Company] Follows Deming Like a Lemming*.[2] No mention of quality in *that* title, is there? A member of the benchmarking team would likely have found that article, among others, if she or he had performed the literature search. Another lesson, of course, is that your key word list must be truly definitive.

External Data Collection: "Live" Data Sources

The best data—the data with the "answers"—typically come from living, breathing sources. Figure 7-6 is a list of some of the more common "live" sources of data.

---

[2]A humorous hypothetical example, but the point is real. No member of a benchmarking team, sitting in front of a computer dialed into an on-line database and knowing the objectives of the study, would have missed that article.

Common "Live" Data Sources
• Current employees of your company.
• Customers.
• Industry analysts and other industry experts.
• Distributors, agents, and manufacturers' representatives.
• Suppliers.
• Data-sharing arrangements.
• Current and former employees of target companies.
• Site visits.

Figure 7-6. Common "live" data sources.

The order shown in Fig. 7-6 (top to bottom) typically approximates the optimal sequence in which these sources should be approached when you are seeking data, although it is unlikely that all sources will be tapped in any single study. The reverse order (starting at the bottom of the list and moving up) best approximates the value of data or information to be obtained from these sources. Intuitively this makes sense. The best data are those that usually require a great deal of work to obtain. Literature searches—the easiest sources of data about companies you are benchmarking—usually provide only superficial levels of data. Site visits, however, can provide the absolute clearest understanding of companies you are studying, but site visits are very difficult to arrange and conduct properly.

Figure 7-7, the data collection staircase, depicts the normal course of data collection. We summarize:

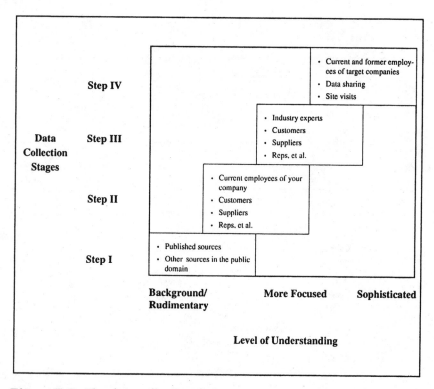

Figure 7-7. The data collection staircase.

- *Step I:* Literature search; reading; beginning to learn.

- *Step II:* Early conversations with people who know more than you do. Typically interspersed with what you will later consider smart or dumb questions you asked. Of course, you probably will not realize just how smart or dumb you were when asking them.

- *Step III:* Good conversations with people who still know more than you, but you're getting smarter. You are learning a lot, and the inane or useless questions pop up less frequently.

- *Step IV:* You are putting the icing on the cake with your interviewing and are feeling a growing, but not yet reliable sense of expertise about your level of knowledge. If you have arranged site visits or data-sharing exercises, you can really learn something. Be sure to avoid reverting to the second step in any site visits you have arranged. At this point, your questions should be as incisive as you can make them.

Let's look at each of the unpublished sources from Fig. 7-6 to understand what they have to offer and how to get it.

Current Employees of Your Company

It is amazing how much your employees know about your competitors, and sometimes about relevant companies outside the industry. A benchmarking team is remiss if internal employees are not among the first to be interviewed. Further, returning to these people later to verify data you obtain or to fill in the pieces of questions you did not have enough knowledge to ask earlier is usually a good idea.

Your human resources department can help you locate employees who used to work for your benchmarking targets. Be sure not to violate any nondisclosure agreements that these people may have with their former employers. Ask your legal counsel if you have any doubts.

Your sales and service people probably encounter your competitors daily and so are a great knowledge base. By the way, if your sales and service people *cannot* tell you much about your competitors, it is time to rethink their job descriptions. They are often the front line of your company and should be gathering data on competitors continually.

Customers

Your benchmarking target's customers will be quite knowledgeable about the following:

- Prices
- Distribution channels
- Promotions
- Product and/or service quality
- Product and/or service comparisons among industry competitors
- After-sales service
- Sales pitches

It is usually best to make these interviews appear to be standard operating procedure, unless you do not care whether your competitors learn of your inquiries. Smart customers will tell you all you want to know if they see their "sharing" to provide them an advantage. These customers may also tell your competitors all *they* want to know. For purposes of a benchmarking study, it is often advisable to have a third party conduct customer interviews for you, so that customers are not made aware that your company is the one interested in the data.

Industry Analysts

Industry analysts generally should be knowledgeable about the following topics:

- Individual company strategies for competing in their industries and the strengths and weaknesses of each
- Strategic groups within industries

- Forces affecting industries and their participants, past, present, and future
- Financial characteristics of industry participants and their relationships with company strategies
- Future plans of industry participants

This widely diverse group of people can be very helpful and includes those who work for trade associations, academicians (if you can reach them[3]), and Wall Street-type investment analysts. This last group can provide varying degrees of helpfulness depending largely on your status as a client, their status in the industry, and the benefits to be gained through cooperation. Some companies become clients of various Wall Street firms merely to have access to Wall Street analysts.

Distributors and Agents

Distributors and agents are generally knowledgeable about the following topics:

- Channel structure
- Channel compensation practices
- Company product and/or service strengths and weaknesses compared to those of competitors
- Channel conflicts
- Trends in evolution of channel structure

Distributers and agents may be able to supply a wealth of information, but they may also be difficult to track down, especially if they travel extensively and they realize, when they speak with you, that you are not buying something from them. You should expect these people, like customers, to pass the word along to your benchmarking targets that you are studying them.

---

[3]Remember, the best academicians are usually very busy people and may be consulting for your competitors. Caveat emptor.

You may want to consider enlisting outside help to interview agents and distributors.

Vendors

Vendors of target companies in your benchmarking study can provide information about the following:

- Costs and usage of materials and components
- Quality and yields of materials and components
- Substitutes for materials and components and their costs and benefits
- Trends in costs and quality
- Inventory levels and composition
- Payment terms and practices
- Supply channels and channel characteristics

Like customers and distributors, vendors can provide in-depth information about target companies. Vendor data should be scrutinized to ensure that what you are hearing is free of sales-pitch hyperbole. After all, vendors would like to sell you 100 percent of your purchased inputs. For that reason, and because vendors, too, tend to talk with your competitors, you may wish to consider having a third party interview vendors also.

Data-Sharing Arrangements

You can obtain some data through data-sharing arrangements among groups of companies that desire to compare their performance within an industry or across different industries. Within industries, data exchanges are usually administered by an independent third party that gathers, compiles, and often analyzes and interprets the data. Figure 7-8 is a sample page from a data exchange among various companies in the life insurance industry. The participants' identities may or may not be revealed in an intraindustry data exchange depending on the level of detail exchanged and the competitive sensitivity of the data. If anonymity is a condition of participation, participants have

Data Point: Current Year First-Year Life Premiums ($000)

Company	First-Year Premium Amount	F/T ORD	Home	Brokers	PPGA	Multi-Line	Mail
A	87,400	58,267			29,133		
B	8,620						8,620
C	387,445	236,341		38,745		112,359	
D	73,415		73,415				
E	98,177	73,632			24,544		
F	203,763	163,010		40,753			
G	11,563						11,563
•							
•							
•							
Z	82,596		78,466	4,130			

Figure 7-8. Data exchange example.

every right to receive written assurance from the independent firm conducting the study that their identities will be kept confidential. Data exchanges, however, many of which are presently being conducted by the Big Six accounting firms and consulting firms that specialize in various industries (e.g., LIMRA in insurance or Hay in compensation), often do not include the qualitative information about where a firm stands in relation to its competitors in a given area. This is the *how* that makes benchmarking so valuable. For this reason data exchanges are seldom the end of a benchmarking effort.

We have seen the entry of Big Six management consultants into "benchmarking." Their provision of sufficient qualitative information in a data-sharing arrangement is sometimes contrary to the relationship these firms typically enjoy with data-sharing participants. A Big Six firm or any national or international firm that undertakes audit or tax work for clients is at risk if it engages, through its management consulting departments, in benchmarking. The appearance of sharing of information among departments is a perception these firms cannot afford. If they do benchmark, the information that the client receives is very limited. These firms often take as many of their clients as they can convince to shell out $XX thousand for marginally helpful data in, say, the electric utility industry, and gather, sort, and report the data. Their objective in these exercises is not usually to provide any one participant with an edge. A more likely objective would be to present the data in such a way as to make as large a subset of the participants as possible, say 90 percent, believe they are in the bottom quartile in at least one area where competitive advantage is necessary, thus causing participants to push the panic button and search for some high-priced consulting help. This may seem cynical. Yet these firms are providing a service which addresses their clients' interests. Until, however, they excuse themselves from proposing on work emerging from their studies, there must be some question as to whose interest is primary.

Across industry lines, data exchanges can be more open as long as participants do not compete with one another in significant markets or as long as the activity being shared is noncore in nature (e.g., in some industries, training activities, MIS, or purchasing). Interindustry data exchanges may be administered by an independent party or by one of the participants, depending on the various factors mentioned above that affect confidentiality.

Current and Former Employees of Target Companies

There is not much you cannot learn from current and former employees of target companies if you spread the net wide enough and are relentless in your pursuit of data. You must have analytical skills good enough to make sense of the many seemingly unrelated pieces of data after the filtering process. When their benchmarking targets are competitors, many companies use outside firms to collect data. Outsiders are often able to do a better and more efficient job of collecting data from their clients' competitors for two reasons. First, they are specialists. They have offices with consultants whose sole job it is to call competitors all day long, looking for that one person in a hundred who can give them the data they need for the "industry study" they are con-

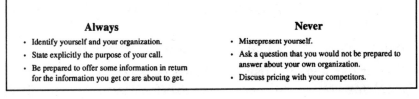

Always	Never
• Identify yourself and your organization.	• Misrepresent yourself.
• State explicitly the purpose of your call.	• Ask a question that you would not be prepared to answer about your own organization.
• Be prepared to offer some information in return for the information you get or are about to get.	• Discuss pricing with your competitors.

Figure 7-9. Ethics in data collection.

ducting. Second, if you were working for Coke and someone from Pepsi called to inquire about the production capacity of your main bottling line, you wouldn't be inclined to answer, would you? And you *must* identify yourself and your organization when calling, for a variety of ethical, if not legal, reasons.

When teaching our benchmarking course, we frequently elicit lively debates among participants about the ethics of competitive benchmarking. Our basic message regarding ethics is included in Fig. 7-9. Beware of the benchmarker who eschews ethics in search of the holy data.

Telephone Interviewing—Some Words of Wisdom. Telephone interviewing is one of the most efficient means of gathering data from target companies; it can also be very frustrating. Interviewing people whom you do not know over the phone requires perseverance, especially in this day of phone mail. Nevertheless, a well-pitched request for a phone interview is one of the most effective ways to gain someone's cooperation in conducting an interview. The ideas below work quite well when they are used by consultancies gathering data over the phone and may also make your phone interviewing efforts go more smoothly.

■ Target the lowest-level person within an organization who might be able to help you first, and move up from there. It is like climbing the data collection staircase: you are learning so you can have higher-level conversations with higher-level people.

■ Always identify yourself and your organization, and give an overview of your project. Try not to give too many details— your potential interviewees may become confused, or you may lead them to believe that what you want is not something they can help with, even though you believe you can learn from them.

- Ask whether you can take just a moment of their time.

- If your party is too busy to talk, ask when it would be convenient for you to call back. Also ask for the names of others with whom you might talk.

- If you get connected to voice mail, don't bother to leave a message. Most people will call you back eventually, but calling you likely is not a high priority for them, and it can be more stressful waiting for them to return your call than continuing to try them. Keep trying to reach them yourself.

- Ask open-ended questions and let your interviewees discuss what they want. You never know where the conversation might take you.

- Be prepared to offer some information as a stimulus to get them to talk. "ABC Company (their competitor) bottles 50,000 bottles per hour and says that's the industry best. How does that sound to you?" Priming the pump like this is a superb way to get people to talk.

Remember, it is human nature to socialize. Exchanging information is a social experience. Skilled interviewers can get answers they need to questions they don't ask simply by engaging in conversation.

Site Visits

Site visits deserve special mention. A well-planned site visit can be the single best way to cap a benchmarking study. No other forum provides the efficiency and ability to dig deeply on a particular study that a site visit does.

But not all studies are best completed with a site visit. In fact, site visits can be a waste of time. In benchmarking a manufacturing process or some other physical activity, a site visit can add considerable value. The flow of product through a plant may be hard to understand without a site visit. Hearing about Federal Express' hub operation—planes landing every 14 seconds, 200,000 packages per night being sorted and delivered, etc.—is not as good as seeing it yourself. But in benchmarking, e.g., managerial processes—training is a perfect example—a site visit may add little value. Whether the added value is sufficient to justify the expense of a trip for the benchmarking team and the time of all those on both sides of the table should be considered carefully before a literal go/no-go decision is made. If all you are going to *see* on a site visit is some faces, ask yourself whether you really have to go. Maybe you could ask the voices at the other end of the phone at the target company to send you their photographs....

Even when a site visit makes sense, it can end up being a boondoggle. On numerous occasions we have heard of managers "flying out to Motorola" or "going to visit plant X," all under the pretense of benchmarking. Preparation: minimal; benefits from visit: less than optimal. Benchmarking is not "flying out for a site visit" if that site visit means a plant tour, some tire kicking, and dinner. A site visit is the *pièce de résistance* of benchmarking, and one should develop an acute appetite for information beforehand.

To be of real value, site visits should be preceded by a great amount of acquisition and analysis of data, preferably acquired directly from the organization being benchmarked in an agreed-upon format that permits ease of analysis and ensures that the data collected are apples-to-apples with comparable data from the benchmarking team's organization. The site visit itself should focus on those few, very detailed issues that are left unanswered by the data alone. The visit should include the line managers of the areas being benchmarked so they can see first-hand that improvement is possible. There is some inherent dissonance in asking an organization that one is imposing on already to prepare rigorously for a site visit. It is like dropping in on a friend and asking him or her to make you dinner. But without rigorous previsit analysis by the benchmarking team, a site visit is not benchmarking at all. Show-and-tell would be a more apt description. Benchmarking of the show-and-tell variety is likely to lose its appeal fast.

One of the most difficult and sensitive problems of site visits is obtaining a level of cooperation from host management that will enable the benchmarking team to maximize its learning opportunity. Three issues are critical and must be addressed when you are contemplating a site visit:

Timing. Host management will need time to prepare for the benchmarking team's visit, which often includes obtaining approval from the executive suite or home office; arranging the schedules of all those who will be involved in the visit; collecting, packaging, and shipping a previsit data package; and preparing the site presentation. It is extremely rare to have a convergence of the stars that allows all the above to happen within a week or two. Remember, host management usually does not have the same incentive or yearning for knowledge that the benchmarking team has. Patience and gentle persuasion are valuable virtues when you are arranging a site visit with a group of managers you have never met.

Timing is often a critical issue when a single site is being visited. The complexity of timing increases exponentially when several sites are being visited. Flexibility and keen logistical skills are keys to smooth success in planning and executing multiple site visits. Your benchmarking consultant is experienced in proposing and negotiating arrangements that meet everyone's interests.

Confidentiality

- *Yours.* News travels quickly. Not long after the plant manager at plant X of ABC Company gets a call from the plant manager at plant Y of DEF Company, requesting a site visit to benchmark the line changeover process, people in the industry may begin to hear two things. (1) Plan X is good at the line changeover process, and (2) plant Y is not. If plant X management does not wish to participate in the study, plant Y management has made known that it wants to improve its line changeover process. Plant Y management received not a single benefit for its trouble. The benchmarking team should consider this issue before deciding who contacts whom.

  If confidentiality is an issue, an artfully crafted letter from an outside firm that is coordinating the benchmarking process can often be used to gauge a target company's willingness to participate. After positive interest has been expressed by the target organization, the outside firm may disclose the benchmarking team's affiliation. This avoids the disclosure-without-results scenario described immediately above.

But there is much debate about the issue of who is most likely to elicit a positive response from a benchmarking target. Where confidentiality is not paramount, initial contact with target companies is often best made by a member of the benchmarking team, not a consultant. Peer-to-peer contact usually provides the best results, since the benchmarking team's counterparts at the target company are going to be doing the work anyway. Using this approach at the outset may help save time.

- *Theirs.* The benchmarking team should be prepared to execute a confidentiality agreement of some sort to ensure that sensitive data from the target organization are not misused. All members of the team must be well informed of the details of the agreement.

Quid pro quo. This is commonly known from the host company's point of view as "What's in it for me?" When several sites from different organizations are being visited, an offer to prepare a compilation of data to be distributed to each participant in the study is often sufficient to obtain participation from target sites. If ensuring that the target plants' identities remain undisclosed among all participants is imperative, this must be adhered to without exception.

To avoid later misunderstandings, these issues should be addressed at the outset of any communications with potential benchmarking partners.

There are other sources of live data, but those just covered are the most common and are used in most benchmarking studies. Before turning to data analysis, consider a couple more points about data collection.

Making Contact with Potential Target Companies

The task that seems to cause the greatest angst in any benchmarking study is the contacting of a real human being at a potential target company that you would like to have join a cooperative or collaborative benchmarking study. It's like asking somebody on a first date.

You can increase the likelihood of getting potential targets to participate in your study by doing a few simple things. There are no guarantees, of course. Some organizations have a policy against sharing, some managers are too busy, and others just do not want to be bothered. On the other side of the coin, some who might be willing participants may be turned off if you send out the wrong signals. The following are some points to consider regarding making first contact with potential targets.

Whom to contact.   As mentioned briefly above, peer to peer usually has the best results, plant manager to plant manager, vice president of sales to vice president of sales, etc. Some companies funnel all requests through a central benchmarking function, others do not. Your peer ultimately will make the decision, though, and may push to participate in a study that someone else in the organization may not consider as important. Selling the idea to your peer is imperative.

How to make contact.   Here the best results are usually obtained by using a two-step approach: a letter followed by a phone call. The letter should make the following points:

- You chose them as a target for the area you are studying because your initial data collection suggested that they are a leader in the field.
- They, too, will benefit from the study (but do not make this claim unless they will).
- They will be expected to contribute some time and data to the study.
- Your benchmarking is going to be intellectually rigorous and not of the industrial tourism variety.
- You will call them on a specific day to answer any questions and, you hope, secure their participation in the study.

The letter is best sent via overnight delivery so it catches the recipient's attention on arrival and you are sure that it got there on time. Then, if you said you would call on Thursday, be sure to do so. Before calling, brainstorm with the benchmarking team to develop a list of potential questions that the recipient may have for you and appropriate responses to each. Then make the call(s). And...

Anticipate some rejection. It is unlikely that every organization you invite to participate in your study will accept. And of those that do, some may not fulfill their obligations. That's business. Don't take it personally. Expect some attrition, and you will be able to maintain a positive, "onward and upward" attitude.

The most important thing you can convey to potential target organizations is your sincerity and intention to complete a value-adding study. The best-in-class companies, which are the most frequent recipients of benchmarking requests, have learned to weed out the serious requests from the industrial tours. Disguising an industrial tour as real benchmarking may get you in the door, but you will have to undress in the foyer and it will not be a pretty sight.

Keeping the Fire Burning— Progress Reports or Other Communiques

Data collection and subsequent analysis take the majority of time in any benchmarking study. It is important to keep interest alive during the data collection period, which can seem interminable for those not actively involved in the process.

Preparing and distributing weekly progress reports is a good way to "keep the fire hot" during the data collection period. The following are some guidelines that we find useful when preparing these reports:

- Distribute the reports only to members of the benchmarking team. Distribution to others in the organization can be an open invitation to unwanted kibitzing and early drawing of half-baked conclusions based on preliminary and incomplete data.

- Include information about the following:

  Who is collecting the data on each of the target companies
  Where each collection effort stands, by target company

Data still needed, by company

Assistance needed, if any, to complete the data collection on
time

Big findings to date (briefly)

- Do not draw any conclusions yourself from the data until
  your collection and analysis efforts are complete.

A Final Word: Data
Collection—Science and Art

Data collection is a mix of science and art. It is science in the
sense that you need to know where to look for data. You need
also to use great care in planning their acquisition to ensure that
all the data needed for a benchmarking study are obtained on
time. Data will not be useful to the team if they arrive outside
the team's specified time frame. Plotting the discovery and capture of data requires that you know where to look and that you
look rigorously.

Actually obtaining data, however, can be quite an art, especially when the information you seek is locked in somebody
else's mind. These data (the data gleaned from many interviews
with people knowledgeable about the benchmarking topic) are
usually the most valuable pieces of a benchmarking jigsaw puzzle. They may be the most difficult pieces to obtain, especially
when the data concern your competitors. You need to know
how to ask, once you have identified the appropriate person to
address.

Envision managers of a beverage distributor who wish to
improve the *cases per stop* of their delivery operation. Stops can
be viewed as fixed costs in the beverage delivery process. Once
the driver arrives at the delivery site, he must find parking, stop
the truck, get out, open the truck bays, find the correct cases of
beverage, load them on the hand truck, deliver them, and get
the retailer to sign a receipt. Many of the tasks in these stops will
take the same time whether the driver is there to deliver 20 cases
or 200. If the driver delivers 200 cases, these "fixed costs" of
finding parking, parking, getting out, opening the truck bays,

and completing the paperwork are spread over many more units, and thus the economics of the process improve significantly. The benchmarking team members wish to learn the cases per stop of some other distributors in the area because they believe their own figure is below average. More importantly, the benchmarkers want to know why others are better, i.e., *how the other companies do it*. So, does the benchmarking team just pick up the phone, call, and ask? Probably not.

In a truly competitive situation, it is likely that an outside consultant will do the data gathering, which will span the gamut of data sources and collection techniques. Outside consultants usually have a team of people assigned to data collection because it can be a grueling task. Searching for the one person who feels like talking on a given day is often frustrating. Once that person has been found, the art of the interview begins. A couple of rules apply:

- *Be prepared.*   You may never get the opportunity to speak again with this one source, who is knowledgeable, ready, and talkative.

- *Never misrepresent yourself.*   Not only is it bad business to do so, but also it is probably illegal if you obtain some solid data that you would not have been able to obtain without misrepresenting yourself.

- *Give some to get some.*   We return to the beverage distributor example. "We have been talking to some of the area distributors, and we understand that XYZ distributor in the next town gets close to 40 cases per stop on average. They say that's tops in the area. Does that sound right to you? Something about how they route their delivery trucks."

"No way. Those guys aren't even close. *We* average almost 65 cases per stop, because our strategy is to concentrate our selling efforts on the biggest accounts. We usually send our managers out to visit them at least once a week, in addition to daily visits from salespeople, and we cater to their every need. They're our bread and butter. We also limit deliveries to our smaller accounts to once per week, so we are able to deliver more at one time. We work with the smaller accounts to better plan their inventory needs, using a system developed by B.I.P. out of Norfolk."

Pay dirt. The age-old tell-them-somebody's-better-than-they-are-and-get-their-reaction technique. It works every time. Interviewing is an art.

Data Analysis

Data collection and data analysis are not discrete, sequential tasks. In the course of most benchmarking studies, collection and analysis take place concurrently and often have a symbiotic relationship. "Let's see. Our initial analysis of data collected on competitor X indicates X is able to produce twice as many units per hour with half as many employees as we have. If this is correct,

- Competitor X's production machinery may be of greater capacity and efficiency.
- Competitor X's employees may be much more highly skilled or trained.
- Some combination of the two contributing factors noted above is occurring.
- Some altogether different reason may exist."

This type of hypothesis testing is often iterative and requires sound judgment from the benchmarking team. As discussed in Chap. 6, having one or more people from the line area being benchmarked is advisable—here's where they again put their expertise to good use.

If there is one big point about data analysis that you should remember, it is that *answering the question "How?" is often more valuable to your benchmarking efforts than answering the question "How much?" Benchmarking is not just a numbers exercise.*[4] Inexperienced benchmarkers frequently miss the boat on this issue, believing that metrics, metrics, metrics are what benchmarking is all about. Metrics *are* extremely important in bench-

---

[4]Sorry for the redundancy, but this point is too important to not be made more than once. You'll see it again.

marking, especially when competitors are being benchmarked. There, true apples-to-apples comparisons can be made. Comparability makes precise analysis more likely. Even then the question "How?" must be answered. When you are studying noncompetitors, "how much" is often important in order to understand orders of magnitude, but less important than "how," due to comparability issues. For example,

- Management of a CAD/CAM software firm benchmarked Xerox, IBM, and Hewlett-Packard to learn how these three provide the world-class after-sales service and support for which they are renowned. Metrics such as the ratios of service and support employees to other groups of employees, the number of calls per day made by field service personnel, and the average response time were very enlightening and much more valuable than just "interesting to know." But the learning about how the target firms structured their after-sales service and support operations, how they motivated and compensated their people, and the many ways in which they used their customer service database was even more valuable to the benchmarking team. This knowledge could be applied directly to the team's own operations. Since the software firm's service process was sufficiently different, the metrics (e.g., number of calls per day) could be used as reliable guides but were not considered as direct apples-to-apples comparisons.
- Managers of the computer hardware manufacturer who discovered that 20 defects per million solder joints were possible (versus the 1500 figure that they frequently experienced) were most interested to learn that simplifying the board layout, reducing variability in the wave solder process, using an expert system for wave soldering, and comprehensive training of employees who worked on the line were factors that contributed to the results achieved by the best-in-class companies.

And, of course, in a strategy benchmarking study (see Chap. 9), "How?" is practically the only question that you want to answer after satisfying yourself, usually through review of financial results or some other indicator of industry leadership, that the companies being studied employ strategies that enable them to achieve superior results.

Common Denominators

Before you can really begin to analyze the data you have collected and turn them into useful information—information that will tell you "how," you have to reduce data from your target companies and from your own organization to common denominators, or put them on an equal basis.

Figure 7-10 shows some data collected by a ready-mix concrete distributor about its major competitors. In this format, these data tell the analyst little. As you look at them, you probably begin to instinctively reduce them to a least common denominator, which is what the benchmarking team did (see Fig. 7-11). These measures were the most meaningful to the company. Note that the team also identified the foremost competitor in each measure and calculated the gap between this distributor and the foremost competitors.

At this point, the benchmarking team has answered the question "How much?" and has likely collected most of the data needed—both explicitly and intuitively—to begin addressing the question "How?"

The benchmarking team's analysis showed that the company could more than close the 1465-yard gap by making two changes: (1) focusing efforts on winning jobs close to plant sites and (2) replacing some smaller (8-cubic-yard) trucks with larger (12-cubic-yard) ones. Closing this gap was instrumental in increasing the company's profitability dramatically in a short period.

| | Company | Competitors | | | |
		A	B	C	D
Estimated yards of concrete poured*	934,000	750,000	500,000	1,000,000	300,000
Number of hourly employees	280	235	160	260	100
Number of salaried employees	45	31	26	40	11
Number of salesmen	14	12	7	14	14
Number of trucks	150	110	85	130	55
Annual maintenance expense estimate*	$810,000	$640,000	$500,000	$775,000	$350,000
Annual depreciation expense estimate*		$1,200,000	$950,000	$1,600,000	$500,000

* These estimates represent the results of extensive data collection efforts.

Figure 7-10. Raw data.

Key Factors	Company Performance	Foremost Competitor		Difference	
		Identity	Amount	Amount	Percent
Annual yardage per truck	6,227	C	7,692	1,465 yards	24%
Annual yardage per hourly employee	3,336	C	3,846	510 yards	15%
Annual yardage per salaried employee	20,756	D	27,273	6,517 yards	31%
Annual yardage per salesman	66,714	D	75,000	8,286 yards	12%
Annual mainte-nance cost per truck	$5,400	B	$5,882	<$482>	<8%>
Annual depre-ciation cost per truck	$12,867	D	$9,090	$3,777	29%

Figure 7-11. Data reduced to common denominators.

Not all competitive gaps can be closed as easily or as quickly, however. It is important to examine gaps by using the old cost accounting technique of analyzing variances to determine which were controllable variances and which were not. In the long run, *all* variances are controllable, of course.[5] But organizational constraints often force the benchmarking team to take a reality-based view of improvements instead of a *Fantasia*-based view. This is discussed further in Chap. 8, which covers implementing benchmarking-driven improvement.

Figure 7-12 shows the benchmarking team's analysis of the various gaps, broken down by controllable and noncontrollable gaps.

Note that management considered the gap in depreciation expense entirely noncontrollable.The benchmarking team

[5]In the long run, we are all dead, too, as John Maynard Keynes once noted.

Measure	Difference Amount / Percent	Controllable Gap	Non-Controllable Gap	Potential Levers to Pull
Annual yardage per truck	1,465 yards / 24%	1600 yards	N/A	• Customer site location • Smaller capacities
Annual yardage per hourly employee	510 yards / 15%	400 yards	110 yards	• Low labor productivity • Poor training of hourly employees
Annual yardage per salaried employee	6,517 yards / 31%	4,500 yards	2,016 yards	• Low labor productivity • Poor office work flow
Annual yardage per salesman	8,286 yards / 12%	10,000 yards	<1,714> yards	• Low labor productivity
Annual maintenance cost per truck	<$482> / <8%>	N/A		_____
Annual depreciation cost per truck	$3,777 / 29%	$0	$3,777	_____

Figure 7-12. Controllable and noncontrollable gaps.

decided that a change in accounting policies or estimates (i.e., method of depreciation or useful lives) was not an avenue it desired to pursue. The company's depreciation policy was conservative and resulted in a higher quality of earnings than a less conservative depreciation policy would have. The economic effect of a change in depreciation policy would have been zero in terms of real performance.

Analysis of competitors' income statements almost always requires close scrutiny of accounting policies to ensure that comparisons are made on an apples-to-apples basis. Depreciation expense and overhead items such as general, selling, and administrative expenses are almost always calculated differently or comprise different elements of a company's true costs.

To reiterate the big point here, note that in answering the question "How?" the benchmarking team began to extract the real benefits from the benchmarking process. The numbers alone tell only part of the story. The managerial processes behind the numbers are the rest of the story, and often the more valuable part.

8

Implementing Improvements

Finish the job.   G. GORDON LIDDY

Introduction

You have completed your data acquisition and analysis and put together a nice report detailing how your organization compares to competitors or other best-in-class companies in the activity you have benchmarked. Unless you do something at this point, however, you have performed a great intellectual exercise with little or no value added. This is the time for action, but your benchmarking study can die right here if you and your teammates don't seize the initiative and push for beneficial change. It is at this juncture in the study that the team's action or lack of it determines whether the study will be a success or not. If changes are not made, if the team does not become the catalyst to make things happen, your benchmarking might turn out to be a waste of time.

This chapter does not intend to put you and your colleagues at the leading edge of change skills. Volumes have been written about that topic, and there is no one correct way to effect change in an organization. There are so many variables at work in change management (culture, inertia, and politics, to name just three) that you and your colleagues can only create an effective plan for change by knowing how those variables interact in your organization. This chapter will help you consider some of the factors that may play a role in determining the eventual success of your entire benchmarking endeavor.

This chapter discusses the bottom three steps in the benchmarking process (see Fig. 8-1).

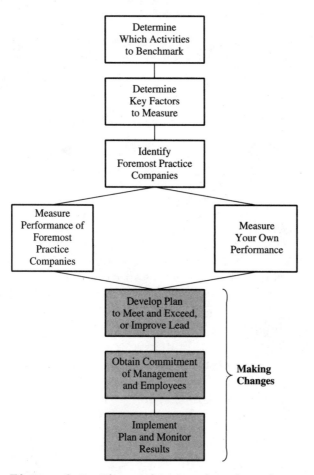

Figure 8-1. The eight-step benchmarking process.

Developing Your Improvement Plan— Determining What to Change

If you have planned your study properly and performed intellectually honest analysis of the data, the changes that your organization should make to improve the activities studied should be clear. Developing a plan to meet or exceed the competition or to fortify an existing lead begins with an analysis of target companies relative to your company. As Robert Camp and his colleagues from Xerox were analyzing L.L. Bean's warehousing and distribution processes, the actions they needed to take to improve Xerox's warehousing and distribution operations became very apparent. Your studies will likely lead you to a similar point. You will know where and how much you can improve. The challenge is to figure out how to make it happen.

We continue with the example on ready-mix concrete from Chap. 7. The company developed a series of tactical moves to close the gap in the areas that were controllable. Most important was the replacement of some smaller-capacity mixers with larger-capacity ones, a move that enabled the company to improve substantially its per-truck yardage. To improve employee productivity, the company developed a basic training curriculum for all employees and implemented a new employee evaluation process—the first such process in the company's history. A summary of the company's planned actions is shown in Fig. 8-2.

Reality Considerations

One of the biggest dangers for a benchmarking team can be that it tries to do too much at once. Benchmarking can be a very rich learning experience, and you may be tempted to try to improve too many things at the same time, running the risk of making marginal improvements in many different areas but not making substantial improvements in the most key areas. Or a benchmarking group may prescribe actions in the key areas that are simply not feasible, given an organization's resources.

Benchmark Measure	Planned Actions
Annual yardage per truck	• Focus attention on large pours close to Company plants • Sell 16 8-yard trucks and replace them with 10 12-yard
Annual yardage per hourly employee	• Allow attrition to reduce headcount. • Begin employee evaluation process. • Provide fundamental training to all hourly employees
Annual yardage per salaried employee	• Streamline office work flow. • Allow attrition to reduce headcount. • Begin employee evaluation process.
Annual yardage per salesman	• Reassign 2 salesmen. • Establish sales quotas. • Begin employee evaluation process.
Annual maintenance cost per truck	Favorable performance
Annual depreciation cost per truck	Noncontrollable
Net improvement	• Cash flow from truck sales and purchases substantially break-even • Annual labor savings of $1.4M • One-time costs: Severance $300K; efficiency study $75K; performance evaluation plan development $50K.

Figure 8-2. Planned actions.

Reality plays a significant role in the implementation of improvements.

Resource limitations, a frequent manifestation of reality, come in many forms. Three of the most common are financial limitations, human limitations, and time limitations. The following are examples of each.

Financial

Managers from a coal-fired electric utility plant benchmarked some of the most efficient comparable plants around the country and learned that their plant could improve its operating performance (e.g., in the areas of heat rate, capacity factor, and forced outage occurrences) dramatically by converting its boilers from pressurized firing to balanced draft. The pressurized-firing to balanced-draft boiler conversion could be done for a total cost of approximately $60 million. Such a large capital expenditure was not in the company's budget for that year. With capital allocation issues and limited resources, it looked as if other projects would take precedence in the next year, too. During the study, the benchmarking team also learned of other improvements that would cost far less than the boiler conversion. Many of these improvements, such as converting boiler controls from pneumatic to electronic, using predictive instead of preventive maintenance in certain areas, and improving fuel mixing techniques, could be made immediately and generate substantial earnings.

When improvements suggested by benchmarking studies include significant capital expenditure projects, remember that the analysis of these improvements must include a rigorous financial analysis, preferably the theoretically correct net present value (NPV) analysis.[1] Those who allocate capital and approve large expenditures always require financial justification for capital outlays and are generally skeptical of intangibles, such as increased customer satisfaction, in lieu of positive NPV. Always run the numbers. Intangibles, like increased customer satisfaction, can usually be quantified in dollars and should be for your analysis to be factual and not appeal strictly to gut feeling.

Human

Human resources—those who can really influence and promote change—are often as scarce as financial resources. Change lead-

---

[1]NPV analysis calculates, as its name suggests, the net present value of a project's cash flows—both into and out of the organization—using the project's cost of capital. For further discussion of how NPV works, see Richard Brealey and Stewart Meyers, *Principles of Corporate Finance*, 3d ed., McGraw-Hill, New York, 1990.

ers are not a dime-a-dozen commodity, and not every organization is blessed with a plethora of change leaders who can make things happen. As discussed in Chap. 6, regarding the composition of a benchmarking team, identifying a change leader is a key task during the planning phase of a study. Take whatever measures you must to have an effective change leader as part of your benchmarking team.

Time

As with the discrete resources of money and people, time is seldom working for us. If time were not a factor and an organization had unlimited human and financial resources, all the "to do" items that are generated by any benchmarking study could be accomplished. But since time is often a reality with which the benchmarking team is forced to cope, a ranking of "to do" items generated by a benchmarking study in their order of relative importance and time requirements is a valuable step in the process.

For example, a multinational professional service firm's headquarters office benchmarked the purchasing functions of Dow Chemical, Florida Power & Light, and Federal Express to see how those companies purchased such things as paper, office supplies, desks, printing, janitorial services, travel services, etc., all inputs that go into making office workers productive. Eventually the headquarters intended to revamp purchasing of all those items, largely by consolidating purchasing power (i.e., buying for regional groups of offices or for the firm's 100+ offices as a whole rather than office by office) and narrowing the vendor base. (Prior to the study, e.g., headquarters bought computer supplies from over 70 vendors. They cut that number to 6 and made plans to "certify" one and an alternate by the following year.) But time constraints kept the relatively small purchasing group of three full-time people from applying these principles at once to all purchased products. The solution they used was simple and has produced good results. They ranked areas of purchasing by dollar volume, which was a proxy for potential savings to the firm, and began to address the areas one at a time. The prioritized list is shown in Fig. 8-3. Travel

Travel services	120
Office supplies and paper	40
Printing	37
Office furniture	24
Temporary services	18
Building and grounds maintenance	12
Computer supplies	8

Figure 8-3. Dollar volume (in millions) of annual purchases.

services obviously came first. On an approximately $120 million annual travel budget, the firm estimated savings of $40 million over the next 3 years. Office supplies and printing were next on the list. By concentrating on the areas with largest potential for improvement first, the firm is allocating its resources effectively to addressing this problem.

Another effective way to prioritize potential improvement actions is to map all the possibilities along a two-dimensional grid with axes chosen from one or more insightful points of view. For example, a travel agent that used benchmarking to determine improvements to be made in the nascent corporate quality process used the two maps shown in Fig. 8-4 to prioritize the changes about to be made. The items that fell into the "hot boxes," the shaded areas, received first priority.

Obtaining Commitment for Change: Political Considerations

If you have planned and executed your benchmarking study properly, you have been building consensus for making changes from the outset. Your benchmarking team includes people from the line who realize that improvements in their processes are necessary, have studied best practices to learn how much to improve, and have learned firsthand that improvement is possible and how it could be done. This involvement of line people in the benchmarking processes is

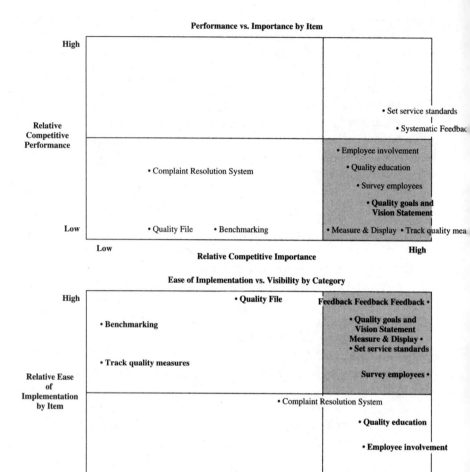

Figure 8-4. Prioritization maps for improving the corporate quality process.

one of the most powerful facets of benchmarking. It can really help minimize the political problems often associated with making improvements in line activities based on recommendations from a headquarters staff person.

It is disheartening to see politics frustrating otherwise bright people who are trying to get something accomplished in corporate America. All organizations comprising three or more people have politics, but some organizations manage to keep the most deleterious effects of politics in check. Others trying to accomplish even the slightest change burn up good time assuaging egos and deferring to people who can hold a beneficial change process hostage. Politics exists, even after benchmarking studies have been planned and executed with the help of line people. Address politics up front. Don't play ostrich.

Figure out who needs to be consulted or considered when making changes as early as you can in any study. Think of all the groups (both inside and outside an organization) that might be affected by any changes that will result from the improvement program you are planning. Those groups that might be affected could potentially impede the change process. They should be considered carefully in the development of any planned improvement actions. Potential groups include all levels of management within your organization, and at a minimum, the line people involved in the change, customers, suppliers, labor unions, etc. Ask yourself whether they are for a change that you propose, indifferent, or against it. Evaluate their support or opposition, both qualitatively and quantitatively. Ask whether and how you can neutralize or convert opposition. Don't run from this type of analysis, especially if your proposed changes add value to your organization.

Competitive Considerations

A large CAD/CAM software vendor benchmarked three large business machines and technology companies that were renowned for their world-class after-sales service and support to gain a sophisticated understanding of what it takes to produce world-class after-sales service and support. Historically in the industry, after-sales service and support was risible. No company stood above the rest in this area. To learn what it took to provide superior service and support, management had to study firms outside the industry. However, management wisely also benchmarked the company's two main competitors to understand precisely their strengths and weaknesses in this area. To ignore competitors would be asking to be caught flat-footed by one or both of them in the event that they were also improving their after-sales service and support operations. The planning, data acquisition, and data analysis phases of the study went extremely well and management of the company was able to arrange site visits with after-sales service and support people from the three world-class companies. Within 6 weeks of starting the study, the benchmarking team had developed detailed planned actions to improve their after-sales service and support in two phases:

- *Phase 1* called for the service and support group to surpass the competition in all important measures within 3 months. These included various response time, staffing, and customer satisfaction measures. Even though big gains were not expected from studying them, competitors were benchmarked because their level of service was considered the absolute floor in level of service the company could offer.

- *Phase 2* called for the service and support group to attain world-class status within 24 months, using processes and metrics from the three world-class companies as starting points.

An important component of the overall planned improvement was to put in place, without great publicity, the infrastructure for delivering improved after-sales service and support. This infrastructure included reorganizing existing service groups, training a cadre of full-time 800-number service center representatives to solve what were very technical problems in one or two phone calls, and building and implementing a sophisticated database to capture all important data about customers, their problems, and the solutions to those problems. Because this process would take some time to complete, only a select group within the company knew of the infrastructure building process and the overall improvement plans in the early stages. This decision was made consciously by senior management. Through benchmarking their two main competitors, top management had learned that one company had begun a serious, but apparently less ambitious, service improvement process of its own. The other company could likely muster the resources to do so quickly if alerted to the fact that the competition was planning organic change within its after-sales service and support function. If either competitor stepped up its after-sales service and support function, it could neutralize any differentiating competitive advantage the company was seeking, which was quite dependent on the "first-mover" advantage management hoped to secure.

Managers developed and executed this strategy because they had first developed a comprehensive competitor response profile, which alerted them to the fact that either competitor could follow suit *and probably would* if it knew that the company was about to change the rules of the game. A competitor response profile (see Chap. 1) is an important component of any benchmarking study. It should be developed during data acquisition and analysis and considered carefully when implementation plans are developed. And it should be developed regardless of whether competitors are among the target firms being benchmarked.

Implementing Your Plan and Monitoring Results

As with any planned action, making benchmarking-driven improvements requires that you include the following features in your plan:

- *Detailed actions* including

  Deadlines and milestones
  Accountability by identified managers for specific goals
  Performance targets that are measurable

- *Scheduled progress reviews* in increments large enough to see results but small enough so people do not forget about the task at hand.

- *Rewards for successful implementation*, linked to performance targets.

- *Plans for contingencies* and corrective actions in the event that things go awry.

- *Plans for periodic recalibration*, which are a competitive necessity in today's world as more and more companies turn to

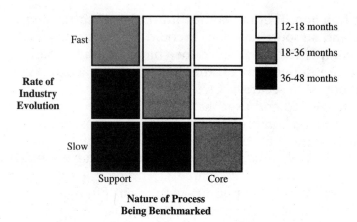

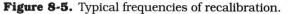

Figure 8-5. Typical frequencies of recalibration.

benchmarking to improve their competitive positions. Frequency of recalibration might best be described as a function of two factors: the rate of industry evolution and the nature of the process being benchmarked. The faster the industry is evolving, the more often you may wish to recalibrate your benchmarks. Similarly, you may wish to recalibrate more frequently those processes that are key to your success, your core functions, and less frequently your support functions. The mix of these two considerations is shown in Fig. 8-5.

Summary of the Benchmarking Process

Using Walter Shewhart's "plan, do, check, act" diagram, we might also view the entire benchmarking process as a cycle like that shown in Fig. 8-6. The improvement process should be continual, like all improvement processes.

Making Improvements— A Final Word

Remember G. Gordon Liddy? He was one of the Watergate players who did his time in jail and now tours the country on

Periodic Recalibration

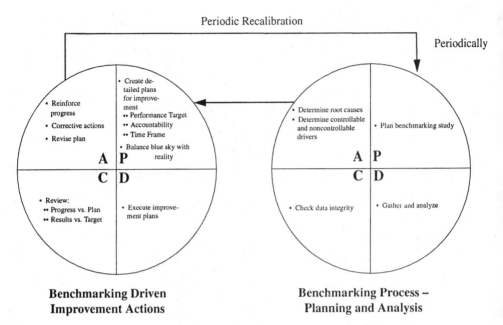

Periodically

Benchmarking Driven   **Benchmarking Process –**
Improvement Actions   **Planning and Analysis**

Figure 8-6. Benchmarking using Shewhart's "plan, do, check, act" diagram.

speaking engagements, proving, once again, that crime does pay.

A recent Liddy quote ends with some apt advice about personal security that might also be applied to the benchmarking process:

When alerted to an intrusion by tinkling glass or otherwise, 1) Arm yourself. 2) Identify the intruder. 3) If hostile, kill him.

Step number 3 is of particular importance. If you leave the guy alive out of misguided softheartedness, he will repay your generosity of spirit by suing you for causing his subsequent paraplegia and seek to force you to support him for the rest of his rotten life. In court he will plead that he was depressed because society had failed him, and he was looking for Mother Teresa for comfort and to offer his services to the poor. In that lawsuit, you will lose. If, on the other hand, you kill him, the most that you can expect is that a relative will bring a wrongful death action. You will have two advantages: First, there will be only your story; forget

Mother Teresa. Second, even if you lose, how much could the bum's life be worth, anyway? A lot less than 50 years worth of paralysis. Don't play George Bush and Saddam Hussein. *Finish the job.*[2]

You have invested a great deal of resources planning your study, gathering data, and analyzing data to learn how much and how you can improve your organization's competitiveness. Now, *finish the job!*

[2]Emphasis added. *The Wall Street Journal,* December 13, 1991, p. 12.

9

Strategy Assessment Using Benchmarking

When your strategy is deep and far-reaching, then what you gain by your calculations is much, so you can win before you even fight. When your strategic thinking is shallow and near-sighted, then what you gain by your calculations is little, so you lose before you do battle. Therefore it is said that victorious warriors win first and then go to war, while defeated warriors go to war first and then seek to win.[1]

SUN TSU

Introduction

The majority of this book has discussed benchmarking as it relates to operations—the execution of strategies that a compa-

[1]No benchmarking book—in fact, few current management books—seems complete without a quote from Sun Tsu, the ancient Chinese general whose book *The Art of War* came into great popularity a few years back.

ny has already developed to compete within its particular industry. The benchmarking methodology is arguably the best methodology for analyzing processes that best-in-class companies use to execute their strategies successfully. But the benchmarking methodology—identify the best; study them; learn from them—can also be applied at a higher level to determine the strategies themselves that are most successful within a particular industry. In other words, if we can use benchmarking to analyze how a firm executes its strategy—how it designs, makes, sells, and services its products—we can use benchmarking to back up a step and determine which strategies are most successful or likely to succeed in the first place. Which are the best firms in a particular industry? Why? What strategies make them so successful?

This chapter outlines a benchmarking methodology that can be used to assess the strategies of participants in practically any industry. You may find this methodology particularly useful when faced with one of the following situations:

- You are considering entry into a new industry
- An industry in which you currently compete is undergoing rapid change
- You are contemplating a shift in strategy within your industry

You may also perform a strategy assessment by using the benchmarking methodology when you simply have never performed a truly rigorous and methodical industry assessment in the past or if those you have done are now dated.

This chapter is not intended to be a definitive work on the multifaceted, deep field of competitive strategy. Thousands of person-years have been used to write hundreds of volumes about the topic, and this chapter merely scratches the surface of it all. Current thinking on the topic of strategy even includes some belief that strategic planning itself is perhaps not as valuable as the ability to seize opportunities as they arise or, conversely, to roll with the punches. Still, it makes sense to use the benchmarking methodology when you are assessing competitors' strategies, so discussion follows of how these types of endeavors are typically approached.

Typical Steps to Strategy Assessment Using Benchmarking

Like any benchmarking project, strategy assessment using benchmarking is not an entirely sequential process. The following steps are presented in an order that roughly approximates the optimal way to approach a strategy assessment study.

Analyze the Overall Industry

This first step is imperative in any type of strategy assessment study. An understanding of the industry structure, its potential for profitability, the state of competition within it, and its likely evolutionary path is the foundation for this piece of the study. Porter's five forces and other frameworks (see Chap. 1) are superb tools for structuring this portion of the study.[2]

Identify the Most Successful Firms

There are a number of ways that success can be measured in identifying firms whose strategies you may wish to study. These measures include

- *Superior financial results.* These include profits, return on sales, return on equity, return on capital employed, etc. When you are identifying firms with superior financial results, it is important to analyze trends over time and to project results into the future. A relatively new measure on the scene has been developed jointly by *The Economist* newspaper and the London Business School (LBS). The measure, intended to provide an indication of a company's quality, is *added value.* Roughly speaking, the measure is calculated by subtracting a capital charge from depreciation-adjusted operating profits. The LBS-*Economist* team found the results to be somewhat out of the mainstream in terms of companies iden-

---

[2]See Michael E. Porter, *Competitive Strategy: Techniques for Analyzing Industries and Competitors,* Free Press, New York, 1980.

1. Autodesk	Computer services	33.9
2. UST	Tobacco	33.7
3. King World	TV, radio	33.0
4. Community Psychiatric	Health care	29.5
5. St. Jude Medical	Medical equipment	29.3
6. Cray Research	Computers	25.6
7. Public Service	Utilities	22.9
8. Merck & Co.	Pharmaceuticals	22.3
9. Eli Lilly	Pharmaceuticals	21.7
10. American Home Products	Pharmaceuticals	21.5

Figure 9-1. Top 10 U.S. firms (added value as percent of sales, 1981/1982 to 1990 average).

tified as best by using this new measure.[3] The top 10 U.S. firms are shown in Fig. 9-1.

- *Superior market share.* Firms with superior market share might not always find their way into the list of firms with superior financial results. But this may be the case because the market leader is posturing for the long run, knowing that higher market share may eventually lead to dominance in the industry. Japanese companies are well known for playing a market share game, and doing it quite successfully. The highest share of market, and the highest cumulative volume that accompanies it, frequently results in a significant cost advantage in an industry due to the effects of the experience curve. Japanese motorcycle and automobile manufacturers are but two examples of groups of companies that have largely exploited the experience curve advantages that accompany high market share. Lester Thurow, in his book *Head to Head* (1992, William Morrow & Company, Inc.), describes the Japanese approach and rationale quite clearly.

- *Growth.* Fast-growing firms might be in neither of the above categories today, but their strategies may be propelling them toward one or both in the future. For that reason alone, they

[3]Copyright © *The Economist*, September 7, 1991, pp. 21–24. Reprinted with permission. Further reproduction prohibited.

should be considered. Another reason that a firm contemplating entry into an industry might wish to look at young, fast-growing firms is that there may be many parallels between the issues those firms have faced in the industry and those you may face during your entrance period.

Analyze the Most Successful Firms Individually

Analyses should encompass, at a minimum, financial results, strategies, and processes in the firms' overall business system.

- *Financial results.* Keeping in mind that accounting differences can skew results and make comparison among firms less meaningful, you should nevertheless be sure to look at some basic financial measures. When possible, look at these measures at the lowest level possible within the target organization.[4]

  Growth. Compound annual growth rates in revenues; earnings before interest and taxes; net income; and assets.

  Liquidity. Current ratio; acid tests; etc.

  Profitability. Return on sales; return on assets; return on equity; and return on capital employed.

  Asset utilization. Day's sales outstanding in receivables; day's cost of goods sold outstanding in inventory and payables; working capital/sales; and fixed assets/sales.

  Leverage. Debt to total capital; debt to equity; debt to total assets; interest coverage; and fixed-charge coverage.

- *Strategies.* Identify each firm's strategy. Is it the overall low-cost producer? If not, how does the firm differentiate its products or services to make buyers perceive higher value and hence pay a higher price?[5]

---

[4]All these ratios are covered in any basic managerial accounting text. For an enlightening discussion of differences in companies' accounting practices (i.e., quality of earnings), see David F. Hawkins, *Corporate Financial Reporting and Analysis*, 3d ed., Irwin, Chicago, 1986, Chap. 9.

[5]See Michael E. Porter, *Competitive Strategy: Techniques for Analyzing Industries and Competitors*, Free Press, New York, 1980, Chap. 2, for a discussion about generic strategies that firms use to compete.

Human Resources (Bring in the people)	Sales and Marketing (Bring in the work)	Products/ Production (Do the work)	Add-on/ Leverage (Sell more work to existing clients or sell similar work to others)	Financial (Results from the work)
• Number • Source • Training • Other	• Key selling points • Process • Promotion • Price • Other	• Types of work performed • Processes employed • Fee structures • Staffing • Alliances	• Technology • Industries • Expertise • Add-on work • Investment	• Revenues • Margins • Profits

Figure 9-2. Simplified business system for information technology industry.

You should be able to articulate clearly, on macro level, each firm's strategy, using just a couple of sentences. The next step in the analysis—breaking down each firm into its business system—will shed a lot more light on your preliminary assessment of each firm's strategy and may cause you to rethink it.

- *Business systems.* Using Porter's value chain or some variation of it that best fits the firms in the industry you are analyzing, break down each firm into its processes. One analyst, studying four successful firms in the broad and fragmented information technology services industry, used the simplified business system shown in Fig. 9-2 to analyze each. While simplistic, the analysis provided fit the benchmarking team's needs on that particular study. A good rule of thumb is to begin with a comprehensive business system like Porter's value chain rather than a simplified version of it. Cutting back on the analysis as you go is generally easier than adding to it.

When you are looking at each firm's business system, be sure to analyze its upstream and downstream linkages. Careful analysis of relationships with suppliers and customers can reveal keys to a strategy that might go unnoticed if the focus is entirely on the internal workings of the firm. The examples in Chap. 6 of the alternatives in the value chain of ready-mix concrete producers provide another illustration of business system analysis and include identification of alternatives in linkages with suppliers.

Compare the Most Successful Firms

Finally, taking the above analyses, now compare all firms against one another. Look for relationships among the various analyses, and ask the two big questions: Why? So what? The following examples are illustrative.

Some Brief Examples of the Importance of Understanding Strategy

In the mid-1970s, England retained one of the leading consultancies to analyze the world motorcycle industry and help devise a strategy for British motorcycle manufacturers to regain profitability and competitive advantage. The most successful firms of the time were the Japanese firms, led by Honda, which had taken the U.S. market by storm in the early 1960s and grown tremendously throughout the 1960s and into the 1970s. A thorough understanding of Honda's strategy in particular was instructive. Honda, through tremendous cumulative volume over time, had built up a significant, sustainable cost advantage due to the effects of the experience curve—so significant and sustainable, in fact, that the British firms eventually exited the industry. Triumphs and BSAs, two of the most popular British motorcycles, today are largely collector's items. Earlier identification of the Japanese firms' strategy may have allowed the British firms to respond successfully. By the time the British figured things out, it was too late to make up the ground they had lost.[6]

Pan Am is another example of a company that did not identify the need for a shift in strategy until it was too late. With the end of airline regulation in the United States in 1978, industry participants were beset with rapid change. In a nutshell, regulation meant firms could survive without worrying too much about being efficient. The government allowed the airlines to price at levels that ensured their profitability.

[6]Harvard Business School Case Services, "Honda" (A), No. 9-384-049, 1983.

When President Carter signed the Airline Deregulation Act in October 1978, plans called for a period of 3 to 5 years before prices were completely free, designed to help incumbent airlines make the changes that would be required to survive in the new, competitive environment. In fact, the entry of almost two dozen new competitors in the first 2 years following deregulation hastened the decline in prices and the strategic moves made by the incumbents to protect their positions in the industry. Specifically, three factors for success in the new environment emerged:

- *Hub-and-spoke systems.*   Instead of the old, point-to-point route systems that the airlines flew, most of the major airlines quickly developed hub-and-spoke systems which gave them increased yields and load factors. United in Chicago, American in Dallas, and Delta in Atlanta are a few examples of airlines that locked up gate space at major airports and built effective hub-and-spoke systems.

- *Computer reservations systems.*   American, United, and TWA developed costly (up to $200 million) computer reservations systems (CRSs) that were tightly integrated into their marketing efforts. These CRSs enabled the airlines to manage costs, yields, and loads with a sophistication that provided tremendous advantages.

- *Low-cost labor.*   Actually, low-cost *happy* labor might be a more accurate way to describe this factor. New entrants, not locked into union agreements or histories of paying high dollars for labor, had obvious advantages here, but the majors mostly made moves to reduce costs in this area. American was the best at reducing labor expense while maintaining a good relationship with employees. This was accomplished largely through a gradual move into two-tiered wages, which paid new employees a more reasonable wage for the services they performed.

Pan Am missed the boat on all three factors. Pan Am considered itself an international airline and never developed a substantial hub-and-spoke system in the United States. Pan Am's fleet—half long-haul widebodies (and old, too—Pan Am was the first customer of the Boeing 747)—was inappropriate for developing short routes in the United States. Pan Am's CRS was a dinosaur, and eventually Pan Am attempted to join American's Sabre system. And Pan Am's labor was high-cost and hostile. Hostile labor results in poor service. Poor service results in fewer passengers. What was labor thinking when it agreed to wage and work-rule concessions in exchange for *equity* in the late 1980s? Don't look up Park Avenue to see the Pan Am building anymore; it's not there.[7]

AT&T provides an interesting study in contrast. In the early 1980s, AT&T decided that its research and development process was so strong that the computer industry was where AT&T wanted to be. If it had been more realistic about the keys to success in the industry, AT&T would have noted that the ability to market, sell, and service personal computers (PCs) was where the winners were separated from the losers. Remember the original AT&T desktop PCs, those slow-as-molasses machines with the green screens that used to blink "Wait" a lot while they were calculating? Some analysts estimate that AT&T poured over $1 billion down the drain to learn that lesson. AT&T should have used that money to help reduce the national debt.

AT&T, though, really has this strategy thing nailed. AT&T's strategy is the best of all known strategies. It's called having lots of cash in the bank. Their next foray into the computer industry came with a turnkey solution. AT&T bought National Cash Register (NCR). NCR kicked all the way down the aisle, but that is what is known as getting a fair price for one's shareholders.

Sarcasm aside, AT&T has done a masterful job at taking one of the world's largest companies and transforming it from a giant monopoly with few worries and organizational skills in competing to a nimble, quality-conscious, customer-driven firm. Two Malcolm Baldrige National Quality Awards went to AT&T in 1992, which says a lot about just how good AT&T's quality is. But winning the Baldrige is not the end of the road for AT&T business units. Neither of those AT&T business units

[7]Harvard Business School Case Services, "The U.S. Airline Industry, 1978–1988" (A) and (B), Nos. 9-390-025 and 9-390-026, 1989.

nor any other AT&T business unit has ever won the chairman's top prize for quality within the AT&T organization.

AT&T's entry into the credit card business is an example of a company understanding the keys for success in an industry and designing a strategy and organization around them. Profits in the bank card industry were high during the 1980s. Most banks charged 17 to 20 percent interest, which was significantly higher than their cost of funds, especially as interest rates declined during the decade.

Able to spot a winner, AT&T decided to enter the industry, but took care to do it correctly. First, AT&T hired Paul Kahn, a long-time veteran of the industry, and a management team comprising other veteran stars from the industry. AT&T's general charge to the team was, "You have been in the industry long enough to know how to put together a winning organization from scratch. Here's the money. Go do it." In an industry where volume is critical, AT&T Universal Card Services (UCS) brought a product to market that was so popular that it became the number two bank card in just 2 years. This large customer base, over 12 million customers at the beginning of 1993, allowed Universal Card Services to spread the large fixed costs of systems and facilities over many units, which brought unit costs down to competitive levels. Since the Universal Card was rewriting the rules of the game by eliminating annual fees, offering lower interest, and providing world-class customer service, low unit costs were critical to the company's success.

To help achieve profitability even faster, managers at UCS devised an ingenious method for bringing in profitable accounts. To understand the genius of their move, it is first necessary to understand some rudimentary economics of the credit card business. Banks that issue credit cards make most of their profits from finance charges, collected on customers' outstanding balances, and annual fees. New customers frequently start off just as you and I do when we get a new credit card. They pay off the entire balance every month. They pledge to themselves that they will not incur any of that costly interest expense. This pledge usually lasts for a few bills. Then the balance begins to creep up. Maybe it's a new TV or a washer-dryer; it could be the holiday season or a wedding anniversary. The next thing you know...

Most banks estimate that it takes 12 to 18 months for an average cardholder's balance to grow large enough for the account to turn profitable. UCS ran a promotion during 1992 to attract new customers to the card by offering inducements to customers who would transfer their balances from other bank cards to the Universal Card. The minimum balance transfer was set high enough that customers signing up for the Universal Card became profitable from day 1.

Clearly AT&T understood the keys for success in the credit card industry. Strategies developed based on a thorough and intellectually sound understanding of the key factors for success in any industry have a much better chance of being implemented successfully than do those based on hunches, gut feelings, or bad information. The benchmarking methodology— identify the best, study them, learn from them, and implement based on that learning—can be applied quite well to addressing strategic issues and should be considered whenever an important strategy issue arises.

10
Benchmarking and the Malcolm Baldrige National Quality Award

Describe the company's processes, current sources and scope, and uses of competitive comparisons and benchmarking information and data to support improvement of quality and overall company operational performance.
1994 AWARD CRITERIA, MALCOLM BALDRIGE NATIONAL QUALITY AWARD

Introduction

The Malcolm Baldrige National Quality Award has played a large role in popularizing benchmarking in corporate America. Much of the early benchmarking being practiced in the United

States was done by the first Baldrige winners and companies they collaborated with on benchmarking studies. As more companies began viewing the Baldrige Award as a worthy corporate goal or the Award's criteria as a good framework for building a system of total quality management (TQM), still more companies were exposed to the benchmarking process. Companies today that are following the Baldrige criteria can't miss the importance of benchmarking—benchmarking or benchmark comparisons are included in six of the seven 1994 Baldrige categories. Only the leadership category makes no mention of benchmarking or benchmarks—you can draw your own conclusion about that.

Talk with your local Baldrige examiner if you have any doubts about how seriously the examiners look at benchmarking when scoring applications or performing site visits. Or talk with past Baldrige winners. They will tell you the same thing. Benchmarking is serious business to the Baldrige people. However, it has also been one of the major areas of weakness found when all Baldrige applications are considered collectively. Many applicants do not benchmark correctly. Let's take a look at what the Baldrige examiners look for in benchmarking when scoring applications or performing site visits.

Overall, the examiners look for methodical and systematic processes related to benchmarking. Remember that most examiners do not review applications or perform site visits for companies in their own industry. The Baldrige's conflict-of-interest rules keep examiners away from companies that compete with their own. An examiner who works for GM, e.g., would not review an application submitted by Ford. And examiners who make their living by consulting will not be reviewing applications from their clients. The end result of these conflict-of-interest rules is that examiners probably will not look at your company while knowing as precisely as you what the appropriate benchmarks are or which company you should be benchmarking. But the examiners will look at your company while knowing what a methodical, systematic benchmarking process is. And they will want to see how you do the following:

- Identify appropriate company processes to be improved through benchmarking.

- Identify appropriate target companies to study.
- Study your target companies.
- Implement what you have learned.

This is the essence of benchmarking, all over again. Let's consider each of these four activities in greater detail.

Identify Appropriate Company Processes to Be Improved through Benchmarking

Having a methodical and systematic way to identify the right processes to benchmark is the first step toward satisfying the Baldrige examiners. The right processes are those processes that have the biggest impact on your business and, hence, on your customers. How do you identify these processes? Chapter 6 covers this issue in greater detail, but if high customer impact and Baldrige examiner satisfaction are your driving forces, make your life simple. Ask your customers what processes you should focus on. While you are at it, ask your competitors' customers, too.

One Baldrige examiner tells of a company that asked all its customers' purchasing agents, "How much additional business would you do with us if we were easy to do business with?" Grammar aside, that's a pretty interesting question. The answer: 60 percent. Next question: "What areas do we most need to improve?" That's called "cutting to the chase" in some circles.

Keep in mind that this approach can deliver the desired results, but examiners want to see that it is done methodically and systematically. Picking up the phone to gather a few data points once or twice will not translate to big Baldrige points. You have to do it through a process that is well thought out and is performed again and again. AT&T Universal Card Services, a 1992 Baldrige winner, does it well. UCS has a Benchmarking Steering Committee that meets periodically to discuss the status of current benchmarking efforts and to determine what to benchmark next. The committee comprises about a dozen representatives from various functions and processes in the company, all of whom are tuned in to the quality of their processes. They achieve this tuned-in state daily through UCS's daily indicators report, which provides data on about 100 critical measures company-wide. Committee members also listen continually to customer feedback, both external and internal, which is collected through several different processes and is fed into the appropriate UCS groups. In short, the UCS Benchmarking Steering Committee is well-informed and customer-focused when making decisions about what to benchmark. Baldrige examiners like that.

Identify Target Companies to Study

To continue our first example from above, after learning from customers what areas it most needed to improve, the company asked another question: "Of all the companies you do business with, who is best at [whatever area the customer named]?" Effective? Probably. Enough? Probably not, although this approach to identifying good benchmarking targets is on the right track.

To satisfy the Baldrige examiners, you should be able to defend your basis for choosing target companies. While competitors may be good companies to target, the fact that they are your competitors may not be, in and of itself, a valid reason for targeting them. Your entire industry may be poor at the process you are trying to improve. The electronic design automation (EDA) software industry provides a good illustration. From the beginning of the industry in the early 1980s through the early 1990s, none of the EDA firms was particularly good at after-sales service and support. Industry competitor A studying industry competitors B and C would be less than forthright to tell Baldrige examiners that those targets were "best" in anything related to after-sales service except, perhaps, passing the buck. Companies in the EDA software industry that wanted to improve their after-sales service and support through benchmarking had to go outside the industry to find anything remotely resembling a bona fide best-in-class company.

Even studying prior Baldrige winners does not give you automatic points with the examiners. For starters, they know where the skeletons are hidden. They also know that winning a Baldrige award does not mean that a company is best at everything. But never exclude past winners from your early potential-targets list. Even if a Baldrige winner is not the best firm to study for a particular area, it is likely that someone there will know which one is.

There is no panacea-like answer to this conundrum, which is among the toughest issues faced by all benchmarkers. The Baldrige examiners look for evidence that the companies you choose to benchmark are leaders and relevant in the area you are trying to improve. Document your case. A recent Baldrige winner trying to improve training benchmarked winners of the American Society of Training and Development (ASTD) National Training Award. *Prima facie,* those companies were leaders in training. Documentation was relatively easy. In other instances, proving that you have benchmarked the right companies may not be so easy. Your goal is to establish that the companies you benchmark are leaders. Independent verification such as the ASTD award just mentioned or other awards given by independent organizations is strong evidence. Hearsay is not. Remember that some companies have earned reputations for being best-in-class at something because they have tremendous marketing and public relations (i.e., self-promoting) skills. Prove to the examiners that the targets you select are the right ones. Compile and document your evidence.

Study the Target Companies

There's a joke about two Baldrige examiners who stumble upon a warm pile on the ground in front of them. While examining the pile, they alternate grunting to each other, "Looks like industrial tourism." "Smells like industrial tourism." "Tastes like industrial tourism." "Hmmm, good thing we didn't step in it."

Baldrige examiners know industrial tourism when they see it. Once they satisfy themselves that your process for identifying areas to study and appropriate target companies is sound, they look to see whether you have truly *studied* your targets or just done some industrial tourism. They also check that you have delved behind the numbers and learned *how* your targets perform the underlying processes that enable them to achieve such outstanding results. To the Baldrige examiners, this understanding of the underlying processes and subsequent improvement of your own processes based on what you have learned is the only way you will be able to make lasting improvements to your quality results (Baldrige award category 6). That is why they look to see what you have learned from your benchmarking. They want to see that your learning is substantial, which means more than just metrics and the platitudes of learning provided by industrial tourism. Substantial learning means understanding details on *how* to make improvements.

Here, too, documenting what you have learned will help you build a case for having done intellectually rigorous and substantial benchmarking. This assumes you have learned enough about the process studied to prepare an instructive document. If you haven't, you have probably not performed what most Baldrige examiners would consider real benchmarking. Save your breath and everybody's time, and offer as benchmarking only those studies you are proud of.

Implement What You Have Learned

Remember, all is for naught if you do not use what you have learned through benchmarking to make improvements in your own organization. Baldrige examiners want to see how you finished the job. They like to see a systematic approach to disseminating findings and deploying process improvements. The existence of teams to deploy improvements is typically viewed favorably. Pilot projects—testing changes on a small scale before rolling them out big-time—are viewed favorably also, where appropriate. The overriding key factor here is to be able to demonstrate that you have actually done something with your benchmarking findings—something more than put them in a binder and put the binder on the shelf.

Preparing for the Baldrige Examiners

This section addresses preparing for a Baldrige examiners' site visit. It does not address writing your Baldrige application. Any attempt to do that would be redundant at best and, more likely, quite presumptuous. The Malcolm Baldrige National Quality Award *Award Criteria* provides the best advice on preparing a Baldrige application. Address the "Areas to Address" in the *Award Criteria* and tell the truth.

In preparing for your site visit, however, you should try to do the following related to benchmarking:

- Have a brief benchmarking presentation ready.
- Overcommunicate your benchmarking message to the rest of your organization.
- Have your documentation handy.
- Expect the unexpected.

Having a brief presentation ready is the corporate benchmarking manifestation of the Boy Scout motto—be prepared.[1] Having your presentation ready will be very helpful if the examiner assigned to visit with you about benchmarking expects one. If not, having prepared one, you have had the opportunity to crystallize your thinking in this area, which is always helpful. Your presentation should cover how your benchmarking is organized, how your company's people are trained in benchmarking, and how you have used benchmarking, following roughly the four activities identified at the outset of this chapter as being important to Baldrige examiners.

Overcommunicating your benchmarking message to others in your organization may be necessary if your benchmarking

---

[1] A fun fact to know and tell: Lord Baden-Powell, who was founder of the Boy Scouts and a hero for the British in the Boer War, coined the Boy Scout motto based on his initials.

endeavors and successes have not been well publicized within the company. Examiners roam around your organization, asking anybody anything related to the Baldrige process. Months before your site visit, you should do the same, to gauge how widespread knowledge about your benchmarking processes is within the company. Those of you who are heavily involved in benchmarking run the risk of taking it for granted that the entire organization is as in tune with benchmarking as you are. Find out for yourself if this is the case before the examiners find out for you. If you discover that most of your organization is not as well versed on your benchmarking endeavors as you would like them to be, start communicating. Write an article or two for the company newsletter, hold some briefing sessions, talk it up at quality meetings. Get the word out!

Having your documentation ready is imperative for your site visit. Be organized. Be able, at a moment's notice, to put your finger on your benchmarking organization chart, the minutes of any benchmarking meetings you may have had, benchmarking training materials, copies of studies you have done and the improvement implementation plans that go with them, correspondence with benchmarking partners, your benchmarking articles file, and, of course, your two benchmarking bibles—Bob Camp's book and this book.

And expect the unexpected. There are hundreds of Baldrige examiners, each with her or his own style, and some revel in the ability to put people on the spot during a site visit. A Baldrige examiner on a site visit once swaggered into a training room like the new sheriff in town, stopped the class dead in its tracks, and started firing questions like two six-shooters at the instructor and the participants. After a couple of minutes, apparently deciding to let the women and children live, he turned and left. Be ready for all kinds of questions, but make sure your examiner asks the questions you want to be asked when you are focusing on benchmarking. You have a story to tell—make sure you tell it. One vice president of benchmarking tells of his experiences with a Baldrige examiner during a mock site visit. The examiner, upon finding out that the vice president grew up in a big college football town and played for the state university, spent almost their entire meeting talking about college football.

The vice president thought the meeting was great. He had come to the meeting not really knowing what to expect and not having any formal agenda prepared. When the mock visit's feedback on benchmarking arrived, though, it was rather poor. It was concluded that the company did not have much of a benchmarking story to tell. Lucky for the company that this had been only a mock site visit. That feedback served as a catalyst to push the benchmarking group to prepare better for the real site visit. It was. And the company won the award.

Benchmarking alone won't do it, but don't let benchmarking hold you back.

11
Using Benchmarking in Your Organization

Bring it home, Sheldon.
UNKNOWN

Introduction

There are as many ways to approach the benchmarking process within an organization as there are organizations performing it, and so many organizations have met with success by using benchmarking that it is quite obvious that there is no one "correct" formula for doing it. Most benchmarking processes fall into one of three approaches: the *training/philosophical* approach, the *ad hoc/grassroots* approach, and the *comprehensive* approach. A summary of these three approaches is included in Fig. 11-1.

The Training/Philosophical Approach

The training/philosophical approach is often used by organizations that wish to instill a sense of competitive awareness in

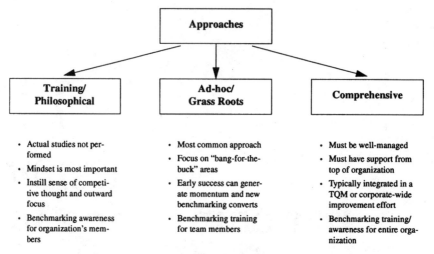

Figure 11-1. Typical corporate benchmarking approaches.

their people. Although benchmarking is not actually performed, understanding the process and philosophy of it is inherently of value. Realizing that the world's creativity and original thought didn't find their genesis within the company's walls can be an epiphany for some organizations. Coming to grips with the benchmarking philosophy—i.e., believing you can learn from others—is often the first step down a path that eventually leads to actual benchmarking.

The main goal of the training/philosophical approach is to get managers and employees to begin to focus on competitors and to realize that there may be radically different ways of doing things—*radically different* meaning better. The telephone and electric utility industries have been awakened to the threat of competition in the past several years. The telephone industry has experienced a two-stage wave of competition, with the long-distance carriers leading the way since the Modified Final Judgment went into effect in 1984. The local telephone companies have only recently begun widespread benchmarking. Both long-distance carriers and locals began benchmarking while taking a philosophical approach. Today, there are still some locals that are easing into real live benchmarking, but most of the Bell operating companies have pushed forward in this area.

Electric utilities, concerned by the threat of cogeneration and bypass, e.g., have also begun benchmarking in earnest. The protection of being regulated monopolies seems to be diminishing, and managers at electric utilities are beginning to behave as managers in competitive industries do. Various consulting groups, including at least one of the Big Six accounting firms, have stepped up to help the electric utility industry move from the philosophical approach to genuine benchmarking.

The Ad Hoc/Grassroots Approach

The ad hoc/grassroots approach is currently the most common approach to benchmarking in corporate America although comprehensive programs are becoming more common. Under the ad hoc approach, managers use benchmarking to help address and improve an area of weakness because they know benchmarking works, not because it is part of an overall organizational push for benchmarking. Managers realize the benefits of the benchmarking methodology and are frequently pioneers of the process in their organizations. When they have finished a study successfully and others learn of the benefits of using benchmarking, the process can take hold at a grassroots level and grow within the organization. The pioneers then become benchmarking gurus at their organizations.

The factors outlined below should be considered when you are planning the initial study. Some of these points have been touched on in earlier chapters but are worth repeating here. Keep in mind that the overriding goal of your study is that it be a success in order to make your organization more competitive. But you also want the study to be a success so it shows others that benchmarking works. And you want it to help you create enthusiasm for benchmarking and provide on-the-job benchmark training for those involved in the project, who can then go off and seed other benchmarking projects.

1. Focus on an area where a little benchmarking is going to provide some big benefits. Chapter 6 discusses this concept in greater detail. An area that most people in your organization

understand or one that has a lot of visibility can also help you when it comes time to communicate how benchmarking helped.

2. Keep a diary about the process that the benchmarking team members go through. This is done in anticipation of other people in your organization asking you to assist in training them. Remember that the dullest pencil is keener than the sharpest memory; take what you learn about the process and write it down.

3. Establish the benchmarking team (see Chap. 6) and "sequester" them as much as possible from the usual hassles that accompany everyday responsibilities, especially during the data collection phase of the study. Gathering data can be extremely challenging and vexing. There are neophyte bench-markers who have balanced their checkbooks, washed their cars, and called their mothers-in-law—something they had not done in years—when they were supposed to be gathering data. The fewer things on one's "to do" list during this phase, the more likely data collection will be completed as originally scheduled. The *ideal* number of things on one's "to do" list during data collection is *one:* data collection.

4. Delineate clearly team members' responsibilities, and set and monitor milestones throughout the study. While never forsaking a positive, can-do attitude, the team leader should have a contingency plan to salvage the study in the event that things do not go as planned. Clearly delineated tasks and frequent monitoring of progress should help minimize the risks of your study crashing and burning.

5. Consider using a consultant—perhaps just as a coach—during your initial study. The benefits are obvious. Benchmarking consultants should be expert at collecting and analyzing data, they can provide unbiased opinions, and usually they do not come with excess baggage in the form of hidden agendas. Using consultants is consistent with the benchmarking philosophy of learning from others, so don't forget to insist on receiving a transfer of their "technology" as part of the deal. Be careful, however, when using consultants that you do not get taken in the process. The mushrooming of benchmarking activity in the United States has caused a corresponding growth in

the number of benchmarking "experts" out there who will be happy to help you—for a fee. Do a thorough background check on them, and be sure that they do what they say they are going to do with the number of people they say they are going to use. Some of these firms prey on the inexperienced benchmarker, who sees the benchmarking process as a black box, and charge an exorbitant fee for mundane work, which they might later try to resell to the competitors. Most firms are reputable, but ask questions before you buy and continue questioning during the engagement.

6. Communicate your success, but wait until you have finished the job. Drawing inferences from incomplete data and communicating them to groups outside the benchmarking team can be embarrassing or worse when subsequent data tell a different story. Novice benchmarkers at an electric utility started comparing one of their plant's operating costs to comparable plants by using cost per megawatthour as reported to the Federal Energy Regulatory Commission (FERC). They discovered that their plant's costs were almost 4 times those of some comparable plants, at least according to the FERC numbers. The FERC numbers, however, are not reported the same by all utilities, which makes apples-to-apples comparisons impossible at that level. Senior management was being "kept abreast" of the study's progress, and the team members were rather embarrassed when they learned through additional data collection and analysis that their plant was closer to 1.5 times more costly, not 4 times; they drew straws to select the bearer of the good-cum-embarrassing news.

7. Communicating your success, of course, means that you must have a solid plan in place after the study to enable you to finish the job. If you have staffed the benchmarking team properly, there is a change leader. Popular methods to promote finishing the job include holding team members accountable for improvements and linking their performance to some sort of incentive scheme.

Communications are sometimes low-key, especially when outside publicity of your study might be competitively perilous. Many benchmarking teams in highly competitive industries—computer hardware and software, to name just two—

walk a fine line between the desire to publicize a tremendous benchmarking success and the need to keep that very fact unknown to competitors.

The Comprehensive Approach

Instituting an organizationwide benchmarking process is a more challenging task than ad hoc benchmarking, but it can pay off tremendously. Recall what companywide benchmarking has done for Xerox, winner of a 1989 Baldrige Award.

Xerox

Almost all Xerox's 100,000+ employees have at least heard of benchmarking, and most know generally what it is all about. For those who want to use the process, or "jump start" benchmarking, according to Robert Camp, Xerox's manager of benchmarking competency, author of *Benchmarking—The Search for Industry Best Practices that Lead to Superior Performance*, and *ex officio* considered the father of benchmarking in the United States, Xerox offers a 2-day formal benchmarking training course. Benchmarking expertise does not reside in one centralized group. Instead, there are a few hundred "competency" people throughout the organization. These competency people serve as benchmarking gurus on teams which are commissioned to perform a study and then decommissioned once the study is complete. These gurus may serve on more than one team at a time, and companywide there may be several dozen studies being performed simultaneously.

Bob Camp's office logs abstracts of completed studies into a database, which is available, with restricted access, to managers across the company. The database is used frequently by managers who wish to find out if work has been done in a particular area before launching a study of their own and by managers who would like information on processes or results that have already been benchmarked by Xerox.

IBM-Rochester

At IBM's 8000+ employee Rochester, Minnesota, manufacturing facility, winner of a 1990 Baldrige Award, benchmarking expertise like Xerox's is decentralized. According to Jerry Balm, the site's senior quality consultant for benchmarking facilitation, the decision to not to have a core benchmarking group was made consciously for two reasons. First, IBM wants its line people to have ownership of any benchmarking that is performed, which means that they budget their own time and other resources for their studies. Second, Balm and his colleagues feel that it is easier to teach a line person how to do benchmarking than to teach a benchmarking person everything he or she needs to know about a particular line process in order to benchmark it effectively. Training is typically done through a one-day course that Balm characterizes as "taking a drink from a fire hose." A great deal of material is covered in this single day, which is the length of time that line managers—Balm's customers—indicated they wanted to spend learning the principles of benchmarking.

Balm heads up a benchmarking council at the Rochester site, which comprises managers from each major group within the division, including hardware development, software development, manufacturing, and site service management. The council meets just two or three times a year to "take stock" of where benchmarking stands within the division. In 1992, the division's focus on benchmarking was to improve the quality of the benchmarking that IBM people performed, so that they would derive increasingly greater benefits from the process and avoid the industrial tourism that may characterize benchmarking at companies that practice it less rigorously.

To monitor the extent of benchmarking being performed, IBM counts *activities*. One process being benchmarked against five target companies equals five "activities." In 1992, over 250 benchmarking activities took place.

Motorola

At Motorola, a 1988 Baldrige winner, benchmarking grew out of the Rise to the Challenge program developed by top management in 1985. The program, of which Motorola's top 3500 managers from around the world were a part, was designed to assist the company in meeting the global competitive challenge that it was facing. Rise to the Challenge pushed managers to use two tools to fortify Motorola's competitiveness—creativity and benchmarking.

Managers at Motorola University in Schaumburg, Illinois, developed a 2½-day course to teach benchmarking skills. Customer demand may force them to shorten the course. The fact that Motorola won the Baldrige Award indicates that there is a lot of benchmarking being done around Motorola's world. But it is all done within the various Motorola organizations and business units, and until recently no central database or group of gurus existed. Motorola is currently rethinking the concept of a centralized database for studies and has developed plans to better capture the results of previously planned studies.

AT&T Universal Card Services

AT&T Universal Card Services (UCS), winner of a 1992 Baldrige Award, has a vice president who oversees the company's benchmarking efforts and leads the UCS benchmarking steering committee. The steering committee comprises about a dozen managers from various groups in the company. They determine collectively which UCS processes to benchmark and oversee progress on various studies. Studies are performed by "process owners," and Universal Card University offers a 2-day course on benchmarking. Results from all studies are "cataloged" and made available to company managers who wish to review studies that have been done before beginning one of their own. Baldrige examiners cited UCS's benchmarking organization as one of the company's strengths in reviewing the company's Baldrige application.

When one encounters comprehensive, organizationwide benchmarking, one usually finds a few other things. First, in most of these organizations, at the top of the company is a belief that benchmarking is a valuable, powerful process. With this belief comes a commitment to support benchmarking, both in work and in deed. This does not usually translate to unlimited resources, but it does mean promoting the process and devoting the necessary resources to enable the organization to get the job done.

Comprehensive benchmarking is also usually accompanied by a centralized repository of benchmarking studies, a central coordinator who oversees benchmarking activities within the company, and a centralized training capability for those who wish to learn how to use the benchmarking process to improve the activity of their choice. Almost by default, in most companies this centralization resides within the "quality group," when such an organization exists. The reason is pretty simple. Benchmarking has advanced so rapidly in the United States largely because the Malcolm Baldrige National Quality Award has promoted and legitimized it. Those charged with Baldrige-related responsibilities usually ended up with benchmarking. This has been terrific for the benchmarking process in general. Quality professionals are usually somewhat fanatic about their jobs, and this fanaticism applied to benchmarking has really helped it mushroom.

Finally, comprehensive benchmarking is usually done in house. Consultants may play a role, especially in training and behind-the-scenes work in data collection for competitive benchmarking studies, but the majority of the work is performed by the organization's own benchmarking-trained people.

No matter which approach they use, all organizations that practice benchmarking have two things in common: a disdain for the status quo and a desire to improve their competitiveness, usually significantly, through a process that is intellectually rigorous and more efficient than most other processes. Corporate America needs to continue this movement to do things better, faster, and cheaper. The huge pool of knowledge that exists can be used to bring our collective competitiveness up a notch or two, and it should be. Benchmarking is a tremendous vehicle for helping us do just that.

12

The Future of Benchmarking

So where is benchmarking going? How is it going to be applied in the United States over the next 5 to 10 years? Is corporate America going to do it well and move our collective competitiveness back up a notch or two?

Benchmarking is being practiced in many different ways, with varying results. As with anything that becomes intellectually fashionable, many people are jumping on the benchmarking bandwagon because it seems like the thing to do. Benchmarking is likely going to help only those who keep the big picture in mind—benchmarking can help improve an organization's competitiveness only if it's done correctly.

This large influx of newcomers to the benchmarking game, often applying their own less rigorous interpretation of the process, has inundated the best-in-class companies with requests to study them. The worst crime against benchmarking—showing up for a site visit without adequate preparation and wasting the host-company's time—has happened frequently enough that many best-in-class firms are making it difficult for the average benchmarker to get an audience. Those that practice poor benchmarking are not doing it intentionally or with malice aforethought. They just don't know better. Establishing one's organization as an organization that practices rigorous, value-adding benchmarking is rapidly becoming one of the prerequisites to learning from the best.

The winners of the Malcolm Baldrige National Quality Award, in particular, are overwhelmed with requests for benchmarking visits. Many of these requests come from organizations that are familiar enough with the award to know that one of the requirements of its winners is to share information on their successful quality strategies with other U.S. companies. However, "sharing information on successful quality strategies" does not necessarily mean honoring every benchmarking request. Managers who are performing real benchmarking, and perhaps have some valuable benchmarks of their own to share, will have the best chance of securing cooperation from the potential target companies of their choice.

Another issue that will become more important is of intercompany communications. It is imperative that benchmarking not be misused to engender cooperation between or among organizations that should be competing. As benchmarking proliferates, companies will form more relationships with other companies that were, prior to benchmarking, their fierce rivals. The sharing of learning and experience that may take place through these interactions is a healthful phenomenon as long as it in no way reduces competition between the sharing companies. Collaborative benchmarking among competitors should be ad hoc, with a clearly defined purpose from the outset, and should be terminated when that purpose has been fulfilled. Collaboration among industry competitors to increase the overall competitiveness of the industry and move the industry generally forward is the ideal outcome of collaborative benchmarking. Anything that reduces overall competition in the industry, however, has the potential to do serious long-term harm and should be avoided.

As expected from any industry that has substantial profit potential, now literally scores of consultancies have sprung up in response to the huge demand for benchmarking coming from corporate America. This has resulted in a broad spectrum of "definitions" of benchmarking processes being proffered by the "experts." The free market will determine the approaches that provide the most value to the organizations using outside consultants.[1] You can bet the ranch that valuable benchmarking will not include completing a series of benchmarking forms, numbers-only exercises, or benchmarking software.

And good benchmarking will not place heavy reliance on benchmarking databases or clearinghouses. Databases of benchmarks are worth little to people serious about using benchmarking for making real improvements because the databases alone will not provide a sufficient level of detail about the underlying processes of the best practices. Those databases that purport to do so will likely only serve as one source in a larger data-gathering effort by companies that are performing serious benchmarking. The power in benchmarking comes from sharing ideas—people sharing ideas—and not from some data in a database of benchmarks or from a library of benchmarking best practices. The real-time live interaction among bright people with good ideas is what makes benchmarking so valuable. The surrealist René Magritte painted a series of pipes, all of which were accompanied by the words "ceci n'est pas une pipe" (this is not a pipe). Applying the same concept to benchmarking, the *forum* of benchmarking is the root of its value, not the benchmarks themselves. Accessing a database or a library is one thing, but it leaves a lot to be desired compared to asking the person who performs the best practice, "Did you ever think about doing it this way?" "Yes, we did, but it didn't work because of such and such." Or "No, we didn't. What a great idea!"

You get the point.

Benchmarking should find its way into the public sector, too, although the lack of competition within the public sector may not create the same sense of urgency to improve that exists in the private sector. Continued swelling of budget deficits might help promote recognition of the need to do things better, faster, and cheaper sometime soon, though, and benchmarking is a natural tool to use for improvement. The opportunities for improvement in the public sector are so vast that benchmarking

---

[1]Unfortunately, this may take some time. The free market is not especially efficient at weeding out value added from smoke and mirrors, particularly because information about benchmarking consultants is rather imperfect.

could have a profound impact on how things are done using our tax dollars. The macroeconomic effects of doing things better, faster, and cheaper in the public sector are frightening, however. There's so much fat that cutting it off is not the answer to the problem. The challenge is to turn the fat into muscle, to make the fat productive. Looking for ways to increase the services we get for the same, or slightly lower, level of spending is an endeavor that can be helped greatly through benchmarking.

If there is one obvious opportunity to use benchmarking to provide some substantial value to our society, it is in the area of public education. Aside from a radical shift in the entire public school paradigm, for which proponents have some very compelling arguments, benchmarking across school systems can add tremendously to the value of our children's education and to the business and administration practices found in today's public education domain. There are literally thousands of school districts across the United States. Some teach science better than others; others, English. Some have a lower dropout rate than others in comparable socioeconomic environs; some have reduced school violence significantly; some have saved many thousands of dollars in their cafeterias; some have student bodies that regularly score above the mean on the Scholastic Aptitude Test. Today, these school districts are not in direct competition with one another. There is no justifiable reason why they should not be sharing their respective accumulated expertise with one another. Benchmarking can help facilitate that process.

Whatever the future of benchmarking, there will be a lot of it being done in the future. Keep in mind that the goal of your benchmarking is to make your organization more competitive, which means that your benchmarking will have to be superior to the next organization's. Remember the bell-shaped curve in Chap. 6. Perform your benchmarking so that if it were judged by a panel of experts, its scores would fall 3 standard deviations to the right of the mean.

Benchmarking is a process, but it also is a way of thinking that is limited only by one's capacity to think. Don't do benchmarking because it's intellectually fashionable. Do it because you are serious about improvement. Do your part to make your organization and this nation's industrial base more competitive.

Appendix A
Selected "Best-in-Class" Organizations

Note: The organizations included in this listing have been noted from a variety of sources including published literature, discussions with industry sources, and benchmarking studies performed by various organizations. The listing is not exhaustive. In addition, the term *best-in-class* has different meanings to different users. Before expending great effort or other resources attempting to study any of the organizations on this list, the benchmarking team should satisfy itself that an organization identified as a potential benchmarking partner is appropriate for study.

Automated Inventory Control

American Hospital
  Supply/Circon Corp.
  460 Ward Dr.
  Santa Barbara, CA 93111-2310
  (805) 967-0404

Apple Computer Inc.
  20525 Mariani Ave.
  Cupertino, CA 95014
  (408) 996-1010

Federal Express Corp.
  2005 Corporate Ave.
  Memphis, TN 38132
  (901) 369-3600

Westinghouse Electric
  Company
  Westinghouse Building,
  Gateway Center
  Pittsburgh, PA 15222
  (412) 244-2000

Benchmarking

AT&T Universal Card Services
8787 Baypine Rd.
Jacksonville, FL 32256
(904) 443-7500

Churchill & Company
P.O. Box 425214
San Francisco, CA 94142
(415) 981-7700

DEC
146 Main St.
Maynard, MA 01754
(508) 493-5111

Florida Power & Light
9250 W. Flagler St.
Miami, FL 33174
(305) 552-3552

Ford
The American Rd.
Dearborn, MI 48121
(313) 322-3000

IBM/Rochester
Highway 52 & 37th St. NW
Rochester, MN 55901
(507) 253-9000

Motorola, Inc.
1303 E. Algonquin Rd.
Schaumburg, IL 60196
(708) 576-5000

Xerox Corp.
800 Long Ridge Rd.
Stamford, CT 06904
(203) 968-3000

Billing and Collection

American Express
World Financial Center
New York, NY 10285
(212) 640-2000

AT&T Universal Card Services
8787 Baypine Rd.
Jacksonville, FL 32256
(904) 443-7500

MCI
1133 19th St. NW
Washington, DC 20036
(202) 872-1600

Bill Scanning

Federal Reserve Bank of New
York
33 Liberty St.
New York, NY 10045-0001
(212) 520-5000

Closed-Loop Customer Feedback

Marriott Corp.
1 Marriott Dr.
Bethesda, MD 20814
(301) 380-9000

Computer After-Sales Service

Hewlett-Packard Co.
300 Hanover St.
Palo Alto, CA 94304
(415) 857-1501

IBM Corp.
Old Orchard Rd.
Armonk, NY 10504
(914) 765-1900

Concurrent Engineering

Boeing Co.
7755 E. Marginal Way S.
Seattle, WA 98108
(206) 655-2121

3M Corporation
3M Center
St. Paul, MN 55144-1000
(612) 733-1110

Customer Focus

GE (plastics)
260 Long Ridge Rd.
Stamford, CT 06927
(203) 357-4000

Wallace Company, Inc.
P.O. Box 1492
Houston, TX 77251-1492
(713) 237-5900

Westinghouse Electric Co. (furniture systems)
Westinghouse Building,
Gateway Center
Pittsburgh, PA 15222
(412) 244-2000

Xerox Corp.
800 Long Ridge Rd.
Stamford, CT 06904
(203) 968-3000

Customer Service

American Express
World Financial Center
New York, NY 10285
(212) 640-2000

AT&T Network Systems Group
Transmission Systems
Business Unit
475 South St.
Morristown, NJ 07962
(201) 606-2000

AT&T Universal Card Services
8787 Baypine Rd.
Jacksonville, FL 32256
(904) 443-7500

Banc One Corp.
100 E. Broad St.
Columbus, OH 43271
(614) 248-5944

Cadillac Motor Car Company
2860 Clark St.
Detroit, MI 48232
(313) 492-7151

Florida Power & Light
9250 W. Flagler St.
Miami, FL 33174
(305) 552-3552

General Electric
260 Long Ridge Rd.
Stamford, CT 06927
(203) 357-4000

Globe Metallurgical, Inc.
6450 Rockside Woods
Blvd. S., #3
Cleveland, OH 44131
(216) 328-0145

Granite Rock Company
P.O. Box 50001
Watsonville, CA 95077
(408) 724-5611

Hewlett-Packard Co.
300 Hanover St.
Palo Alto, CA 94304
(415) 857-1501

IBM Rochester
Highway 52 & 37th St. NW
Rochester, MN 55901
(507) 253-9000

L.L. Bean Inc.
  Casco St.
  Freeport, ME 04033
  (207) 865-4761

Marlow Industries
  10451 Vista Park Rd.
  Dallas, TX 75238-2645
  (214) 340-4900

Milliken & Company
  920 Milliken Rd.
  Spartanburg, SC 29303
  (803) 573-2020

Motorola, Inc.
  1303 E. Algonquin Rd.
  Schaumburg, IL 60196
  (708) 576-5000

Nordstrom Inc.
  1501 5th Ave.
  Seattle, WA 98101-1603
  (206) 628-2111

Procter & Gamble Co.
  1 Procter & Gamble Plaza
  Cincinnati, OH 45202
  (513) 983-1100

Ritz Carlton Hotel Co.
  3414 Peachtree Rd. NE,
    Suite 300
  Atlanta, GA
  (404) 237-5500

Solectron Group
  2001 Fortune Dr.
  San Jose, CA 95131
  (408) 957-8500

Texas Instruments Inc.
  13500 N. Central Expressway
  Dallas, TX 75265
  (214) 995-3333

Wallace Company, Inc.
  P.O. Box 1492
  Houston, TX 77251-1492
  (713) 237-5900

Westinghouse Commercial
    Nuclear Fuel Division
  Westinghouse Building,
    Gateway Center
  Pittsburgh, PA 15222
  (412) 244-2000

Xerox Corp.
  800 Long Ridge Rd.
  Stamford, CT 06904
  (203) 968-3000

Zytec Corp.
  7575 Market Place Dr.
  Eden Prairie, MN 55344
  (612) 941-1100

Customer Survey Techniques

AT&T Universal Card Services
  8787 Baypine Rd.
  Jacksonville, FL 32256
  (904) 443-7500

Marriott Corp.
  1 Marriott Dr.
  Bethesda, MD 20814
  (301) 380-9000

Motorola, Inc.
  1303 E. Algonquin Rd.
  Schaumburg, IL 60196
  (708) 576-5000

Zytec Corp.
  7575 Market Place Dr.
  Eden Prairie, MN 55344
  (612) 941-1100

Data Transfer/Check Clearing

First National Bank of Chicago
1 First National Plaza,
Suite 0518
Chicago, IL 60670
(312) 732-4000

Design for Manufacturing Assembly

Digital Equipment Corporation
146 Main St.
Maynard, MA 01754
(508) 493-5111

Motorola, Inc.
1303 E. Algonquin Rd.
Schaumburg, IL 60196
(708) 576-5000

NCR Corp.
1700 S. Patterson Blvd.
Dayton, OH 45479
(513) 445-5000

Document Processing

Citicorp
399 Park Ave.
New York, NY 10043
(212) 559-1000

Electronic Data Imaging

Bell Atlantic Corp.
1717 Arch St.
Philadelphia, PA 19103
(215) 963-6000

Employee Recognition

AT&T Universal Card Services
8787 Baypine Rd.
Jacksonville, FL 32256
(904) 443-7500

Milliken and Co.
920 Milliken Rd.
Spartanburg, SC 29303
(803) 573-2020

Employee Suggestions

Dow Chemical Co.
2030 Willard H. Dow Center
Midland, MI 48674
(517) 636-1000

Milliken and Co.
920 Milliken Rd.
Spartanburg, SC 29303
(803) 573-2020

Procter & Gamble Co.
1 Procter & Gamble Plaza
Cincinnati, OH 45202
(513) 983-1100

Toyota Motor Manufacturing
U.S.A. Inc.
1001 Cherry Blossom Way
Georgetown, KY 40324
(502) 868-2000

Employee Surveys

Mayflower Group Inc.
9998 N. Michigan Rd.
Carmel, IN 46032
(317) 875-1469

Empowerment

Honda of America
    Manufacturing Inc.
Honda Parkway
Marysville, OH 43040
(513) 642-5000

Milliken and Co.
920 Milliken Rd.
Spartanburg, SC 29303
(803) 573-2020

Environmental Management

3M Corporation
3M Center
St. Paul, MN 55144-1000
(612) 733-1110

Ben & Jerry's Homemade Inc.
Junction of Routes 2 & 100
N. Moretown, VT 05676
(802) 244-5641

Dow Chemical Co.
2030 Willard H. Dow Center
Midland, MI 48674
(517) 636-1000

Executive Development

General Electric
260 Long Ridge Rd.
Stamford, CT 06927
(203) 357-4000

Facilities Management

Walt Disney World Co.
1675 Buena Vista Dr.
Buena Vista, FL 32830
(407) 824-2222

Flexible Manufacturing

Allen-Bradley Company Inc.
1201 S. 2d St.
Milwaukee, WI 53204
(414) 382-2000

Baldor Electric Co.
5711 S. 7th St.
Fort Smith, AK 72902
(501) 646-4711

Motorola/Boynton Beach
1303 E. Algonquin Rd.
Schaumburg, IL 60196
(708) 576-5000

Health Care Management

Allied-Signal Aerospace Co.
2525 W. 190th St.
Torrance, CA 90504
(213) 321-5000

Adolph Coors Co.
12th St. & Ford St.
Golden, CO 80401
(303) 279-6565

Southern California Edison Co.
2244 Walnut Grove
Rosemead, CA 91770
(818) 302-1212

Human Resources Management

Baxter International Inc.
1 Baxter Parkway
Deerfield, IL 60015
(708) 948-2000

Walt Disney World Co.
1675 Buena Vista Dr.
Buena Vista, FL 32830
(407) 824-2222

IBM Corp.
Old Orchard Rd.
Armonk, NY 10504
(914) 765-1900

L.L. Bean Inc.
Casco St.
Freeport, ME 04033
(207) 865-4761

Ritz Carlton Hotel Co.
3414 Peachtree Rd. NE,
Suite 300
Atlanta, GA 30326
(404) 237-5500

3M Corporation
3M Center
St. Paul, MN 55144-1000
(612) 733-1110

Industrial Design

Black & Decker Corp. (house-
hold products)
701 E. Joppa Ave.
Towson, MD 21204
(410) 583-3900

Braun Corp.
1014 S. Monticello St.
Winamac, IN 46996
(219) 946-6153

Herman Miller Inc.
8500 Byron Rd.
Zeeland, MI 49464
(616) 772-3300

Inventory Control

American Hospital
Supply/Circon Corp.
460 Ward Dr.
Santa Barbara, CA 93111-2310
(805) 967-0404

L.L. Bean Inc.
Casco St.
Freeport, ME 04033
(207) 865-4761

Northern Telecom Inc.
200 Athens Way
Nashville, TN 37228
(615) 734-4000

Inventory Management Systems

Northern Telecom Inc.
200 Athens Way
Nashville, TN 37228
(615) 734-4000

Make-versus-Buy Decisions

Boeing Co.
7755 E. Marginal Way S.
Seattle, WA 98108
(206) 655-2121

Manufacturing

AT&T Network Systems Group
Transmission Systems
Business Unit
475 South St.
Morristown, NJ 07962
(201) 606-2000

IBM Corp.
Old Orchard Rd.
Armonk, NY 10504
(914) 765-1900

Hewlett-Packard Co.
300 Hanover St.
Palo Alto, CA 94304
(415) 857-1501

Milliken & Company
920 Milliken Rd.
Spartanburg, SC 29303
(803) 573-2020

Motorola, Inc.
1303 E. Algonquin Rd.
Schaumburg, IL 60196
(708) 576-5000

Solectron Group
2001 Fortune Dr.
San Jose, CA 95131
(408) 957-8500

Texas Instruments Inc.
13500 N. Central Expressway
Dallas, TX 75265
(214) 995-3333

Toyota Motor Manufacturing
    U.S.A. Inc.
1001 Cherry Blossom Way
Georgetown, KY 40324
(502) 868-2000

Westinghouse Commercial
    Nuclear Fuel Division
Westinghouse Building,
    Gateway Center
Pittsburgh, PA 15222
(412) 244-2000

Xerox Corp.
800 Long Ridge Rd.
Stamford, CT 06904
(203) 968-3000

Zytec Corp.
7575 Market Place Dr.
Eden Prairie, MN 55344
(612) 941-1100

Manufacturing Operations Management

Corning Inc.
Houghton Park
Corning, NY 14831
(607) 974-9000

Hewlett-Packard Co.
300 Hanover St.
Palo Alto, CA 94304
(415) 857-1501

Phillip Morris Companies Inc.
120 Park Ave.
New York, NY 10017
(212) 880-5000

Marketing

3M Corporation
3M Center
St. Paul, MN 55144-1000
(612) 733-1110

Helene Curtis Industries Inc.
325 N. Wells St.
Chicago, IL 60610
(312) 661-0222

IBM Corp.
Old Orchard Rd.
Armonk, NY 10504
(914) 765-1900

Microsoft Corp.
1 Microsoft Way
Redmond, WA 98052-6399
(206) 882-8080

Procter & Gamble Co.
1 Procter & Gamble Plaza
Cincinnati, OH 45202
(513) 983-1100

Xerox Corp.
800 Long Ridge Rd.
Stamford, CT 06904
(203) 968-3000

MIS

AMR Corp.
P.O. Box 619616
Dallas–Fort Worth Airport,
TX 75261-9616
(817) 963-1234

Banc One Corp.
100 E. Broad St.
Columbus, OH 43271
(614) 248-5944

Bankers Trust New York Corp.
280 Park Ave.
New York, NY 10017
(212) 250-2500

General Dynamics Corp.
3190 Overview Park Dr.
Falls Church, VA 22042-4523
(703) 876-3000

General Electric
260 Long Ridge Rd.
Stamford, CT 06927
(203) 357-4000

GTE Corp.
1 Stamford Forum
Stamford, CT 06904
(203) 965-2000

MCI
1133 19th St. NW
Washington, DC 20036
(202) 872-1600

Security Pacific Corp.
333 S. Hope St.
Los Angeles, CA 90071
(213) 345-6211

Plant Layout and Design

Cummins Engine Company Inc.
500 Jackson St.
Columbus, IN 47202-3005
(812) 377-5000

General Electric
260 Long Ridge Rd.
Stamford, CT 06927
(203) 357-4000

Policy Deployment

Florida Power & Light
9250 W. Flagler St.
Miami, FL 33174
(305) 552-3552

Ford
The American Rd.
Dearborn, MI 48121
(313) 322-3000

Process Improvement

CSX Corp.
901 E. Cary St.
Richmond, VA 23219
(804) 782-1400

Federal Express Corp.
2005 Corporate Ave.
Memphis, TN 38132
(901) 369-3600

Motorola
1303 E. Algonquin Rd.
Schaumburg, IL 60196
(708) 576-5000

Solectron Group
2001 Fortune Dr.
San Jose, CA 95131
(408) 957-8500

Xerox Corp.
  800 Long Ridge Rd.
  Stamford, CT 06904
  (203) 968-3000

Product Development

American Express
  World Financial Center
  New York, NY 10285
  (212) 640-2000

Digital Equipment Corporation
  146 Main St.
  Maynard, MA 01754
  (508) 493-5111

Federal Express Corp.
  2005 Corporate Ave.
  Memphis, TN 38132
  (901) 369-3600

Hewlett-Packard Co.
  300 Hanover St.
  Palo Alto, CA 94304
  (415) 857-1501

Honda of America
    Manufacturing Inc.
  Honda Parkway
  Marysville, OH 43040
  (513) 642-5000

Intel Corp.
  2200 Mission College Blvd.
  Santa Clara, CA 95052
  (408) 765-8080

3M Corporation
  3M Center
  St. Paul, MN 55144-1000
  (612) 733-1110

Motorola
  1303 E. Algonquin Rd.
  Schaumburg, IL 60196
  (708) 576-5000

Sony Corporation of America
  Sony Dr.
  Park Ridge, NJ 07656
  (201) 930-1000

Product Improvement

Northern Telecom Inc.
  200 Athens Way
  Nashville, TN 37228
  (615) 734-4000

Motorola
  1303 E. Algonquin Rd.
  Schaumburg, IL 60196
  (708) 576-5000

Project Team Management

Chaparral Steel Co.
  300 Ward Rd.
  Midlothian, TX 76065-9651
  (214) 775-8241

Purchasing

AMP Inc.
  471 Friendship Rd.
  Harrisburg, PA 17105
  (717) 564-0100

Federal Express Corp.
  2005 Corporate Ave.
  Memphis, TN 38132
  (901) 369-3600

Florida Power & Light
  9250 W. Flagler St.
  Miami, FL 33174
  (305) 552-3552

Honda of America
  Manufacturing Inc.
Honda Parkway
Marysville, OH 43040
(513) 642-5000

NCR Corp.
1700 S. Patterson Blvd.
Dayton, OH 45479
(513) 445-5000

Xerox Corp.
800 Long Ridge Rd.
Stamford, CT 06904
(203) 968-3000

Quality Management

AT&T Network Systems Group
  Transmission Systems
  Business Unit
475 South St.
Morristown, NJ 07962
(201) 606-2000

AT&T Universal Card Services
8787 Baypine Rd.
Jacksonville, FL 32256
(904) 443-7500

Cadillac Motor Car Company
2860 Clark St.
Detroit, MI 48232
(313) 492-7151

Digital Equipment Corporation
146 Main St.
Maynard, MA 01754
(508) 493-5111

Federal Express Corp.
2005 Corporate Ave.
Memphis, TN 38132
(901) 369-3600

Globe Metallurgical, Inc.
6450 Rockside Woods
  Blvd. S., #3
Cleveland, OH 44131
(216) 328-0145

Granite Rock Company
P.O. Box 50001
Watsonville, CA 95077
(408) 724-5611

IBM Rochester
Highway 52 & 37th St. NW
Rochester, MN 55901
(507) 253-9000

Marlow Industries
10451 Vista Park Rd.
Dallas, TX 75238-2645
(214) 340-4900

Milliken & Company
920 Milliken Rd.
Spartanburg, SC 29303
(803) 573-2020

Motorola, Inc.
1303 E. Algonquin Rd.
Schaumburg, IL 60196
(708) 576-5000

Ritz Carlton Hotel Co.
3414 Peachtree Rd. NE,
  Suite 300
Atlanta, GA
(404) 237-5500

Solectron Group
2001 Fortune Dr.
San Jose, CA 95131
(408) 957-8500

Texas Instruments Inc.
13500 N. Central Expressway
Dallas, TX 75265
(214) 995-3333

Wallace Company, Inc.
P.O. Box 1492
Houston, TX 77251-1492
(713) 237-5900

Westinghouse Commercial
Nuclear Fuel Division
Westinghouse Building,
Gateway Center
Pittsburgh, PA 15222
(412) 244-2000

Xerox Corp.
800 Long Ridge Rd.
Stamford, CT 06904
(203) 968-3000

Zytec Corp.
7575 Market Place Dr.
Eden Prairie, MN 55344
(612) 941-1100

Quality Process

AT&T Network Systems Group
Transmission Systems
Business Unit
475 South St.
Morristown, NJ 07962
(201) 606-2000

AT&T Universal Card Services
8787 Baypine Rd.
Jacksonville, FL 32256
(904) 443-7500

Cadillac Motor Car Company
2860 Clark St.
Detroit, MI 48232
(313) 492-7151

Federal Express Corp.
2005 Corporate Ave.
Memphis, TN 38132
(901) 369-3600

Florida Power & Light
9250 W. Flagler St.
Miami, FL 33174
(305) 552-3552

Globe Metallurgical, Inc.
6450 Rockside Woods
Blvd. S., #3
Cleveland, OH 44131
(216) 328-0145

Granite Rock Company
P.O. Box 50001
Watsonville, CA 95077
(408) 724-5611

IBM Rochester
Highway 52 & 37th St. NW
Rochester, MN 55901
(507) 253-9000

Marlow Industries
10451 Vista Park Rd.
Dallas, TX 75238-2645
(214) 340-4900

Milliken & Company
920 Milliken Rd.
Spartanburg, SC 29303
(803) 573-2020

Motorola, Inc.
1303 E. Algonquin Rd.
Schaumburg, IL 60196
(708) 576-5000

Ritz Carlton Hotel Co.
3414 Peachtree Rd. NE,
Suite 300
Atlanta, GA
(404) 237-5500

Solectron Group
2001 Fortune Dr.
San Jose, CA 95131
(408) 957-8500

Texas Instruments Inc.
13500 N. Central Expressway
Dallas, TX 75265
(214) 995-3333

Toyota Motor Manufacturing
U.S.A. Inc.
1001 Cherry Blossom Way
Georgetown, KY 40324
(502) 868-2000

Wallace Company, Inc.
P.O. Box 1492
Houston, TX 77251-1492
(713) 237-5900

Westinghouse Electric
Company
Westinghouse Building,
Gateway Center
Pittsburgh, PA 15222
(412) 244-2000

Xerox Corp.
800 Long Ridge Rd.
Stamford, CT 06904
(203) 968-3000

Zytec Corp.
7575 Market Place Dr.
Eden Prairie, MN 55344
(612) 941-1100

Quick Changeovers

Dana Corp.
4500 Dorr St.
Toledo, OH 43697
(419) 535-4500

Johnson Controls Inc.
5757 N. Green Bay Ave.
Milwaukee, WI 53201
(414) 228-1200

United Electric Controls Co.
180 Dexter Ave.
Watertown, MA 02172
(617) 926-1000

Research and Development

AT&T Co.
32 Avenue of the Americas
New York, NY 10013-2412
(212) 605-5500

Hewlett-Packard Co.
300 Hanover St.
Palo Alto, CA 94304
(415) 857-1501

Shell Oil Co.
1 Shell Plaza
Houston, TX 77252
(713) 241-6161

Robotics

General Electric
260 Long Ridge Rd.
Stamford, CT 06927
(203) 357-4000

Sales Management

IBM Corp.
Old Orchard Rd.
Armonk, NY 10504
(914) 765-1900

Merck and Company Inc.
P.O. Box 2000
Rahway, NJ 07065-0909
(908) 594-4000

Procter & Gamble Co.
1 Procter & Gamble Plaza
Cincinnati, OH 45202
(513) 983-1100

Xerox Corp.
  800 Long Ridge Rd.
  Stamford, CT 06904
  (203) 968-3000

Self-Directed Work Teams

Corning Inc. (SCC plant)
  Houghton Park
  Corning, NY 14831
  (607) 974-9000

Physio-Control Corp.
  11811 Willows Rd. NE
  Redmond, WA 98052
  (206) 867-4000

Toledo Scale/Mettler
    Instrument Corp.
  Princeton-Hightstown Rd.
  Hightstown, NJ 08520
  (609) 448-3000

Service Parts Logistics

Deere and Co.
  John Deere Rd.
  Moline, IL 61265-8098
  (309) 765-8000

**Supplier Technical
Assistance**

Corning Inc.
  Houghton Park
  Corning, NY 14831
  (607) 974-9000

Globe Metallurgical Inc.
  6450 Rockside Woods
    Blvd. S., #3
  Cleveland, OH 44131
  (216) 328-0145

Xerox Corp.
  800 Long Ridge Rd.
  Stamford, CT 06904
  (203) 968-3000

Supplier Management

Bose Corp.
  The Mountain
  Framingham, MA 01701
  (508) 879-7330

Ford
  The American Rd.
  Dearborn, MI 48121
  (313) 322-3000

Levi Strauss and Co.
  1155 Battery St.
  San Francisco, CA 94120
  (415) 544-6000

Motorola, Inc.
  1303 E. Algonquin Rd.
  Schaumburg, IL 60196
  (708) 576-5000

3M Corporation
  3M Center
  St. Paul, MN 55144-1000
  (612) 733-1110

Xerox Corp.
  800 Long Ridge Rd.
  Stamford, CT 06904
  (203) 968-3000

Surface Mount Technology

Hewlett-Packard Co.
  300 Hanover St.
  Palo Alto, CA 94304
  (415) 857-1501

Fuji-Xerox Film Company Ltd.
555 Taxter Rd.
Elmsford, NY 10523
(914) 789-8100

Motorola, Inc.
1303 E. Algonquin Rd.
Schaumburg, IL 60196
(708) 576-5000

Technology Transfer

Dow Chemical Co.
2030 Willard H. Dow Center
Midland, MI 48674
(517) 636-1000

3M Corporation
3M Center
St. Paul, MN 55144-1000
(612) 733-1110

Square D Co.
1415 S. Roselle
Palatine, IL 60067
(708) 397-2600

Telecommunications

Motorola, Inc.
1303 E. Algonquin Rd.
Schaumburg, IL 60196
(708) 576-5000

Telephony

AT&T Universal Card Services
8787 Baypine Rd.
Jacksonville, FL 32256
(904) 443-7500

Baxter International Inc.
1 Baxter Parkway
Deerfield, IL 60015
(708) 948-2000

USAA
USAA Bldg.
San Antonio, TX 78228
(210) 498-2211

Total Productive Maintenance

Eastman Kodak Co.
Eastman Rd.
Kingsport, TN 37662
(615) 229-2000

Training

AT&T Universal Card Services
8787 Baypine Rd.
Jacksonville, FL 32256
(904) 443-7500

Ford
The American Rd.
Dearborn, MI 48121
(313) 322-3000

General Electric
260 Long Ridge Rd.
Stamford, CT 06927
(203) 357-4000

Motorola, Inc.
1303 E. Algonquin Rd.
Schaumburg, IL 60196
(708) 576-5000

Polaroid Corp.
549 Technology Sq.
Cambridge, MA 02139
(617) 577-2000

Square D Co.
1415 S. Roselle
Palatine, IL 60067
(708) 397-2600

USAA
USAA Bldg.
San Antonio, TX 78228
(210) 498-2211

Wallace Company, Inc.
P.O. Box 1492
Houston, TX 77251-1492
(713) 237-5900

Vendor Certification

Boeing Co.
7755 E. Marginal Way S.
Seattle, WA 98108
(206) 655-2121

Cummins Engine Company Inc.
500 Jackson St.
Columbus, IN 47202-3005
(812) 377-5000

Dow Chemical Co.
2030 Willard H. Dow Center
Midland, MI 48674
(517) 636-1000

Motorola, Inc.
1303 E. Algonquin Rd.
Schaumburg, IL 60196
(708) 576-5000

Warehousing and Distribution

Citicorp
399 Park Ave.
New York, NY 10043
(212) 559-1000

Federal Express Corp.
2005 Corporate Ave.
Memphis, TN 38132
(901) 369-3600

Hershey Foods Corp.
100 Crystal A Dr.
Hershey, PA 17033-0810
(717) 534-4001

L.L. Bean Inc.
Casco St.
Freeport, ME 04033
(207) 865-4761

Mary Kay Cosmetics Inc.
8787 Stemmons Freeway
Dallas, TX 75247
(214) 630-8787

Waste Minimization

Dow Chemical Co.
2030 Willard H. Dow Center
Midland, MI 48674
(517) 636-1000

3M Corporation
3M Center
St. Paul, MN 55144-1000
(612) 733-1110

Industry Associations[1]

Aerospace

IEEE Aerospace and Electronics
   Systems Society (AESS)
   c/o Institute of Electrical and
   Electronics Engineers
345 E. 47th St.
New York, NY 10017
(212) 705-7867

National Aeronautic
   Association of the U.S.A.
   (NAA)
1815 N. Fort Myer Dr.
Arlington, VA 22209
(703) 527-0226

Aerospace Database
Technical Information Service
American Institute of
   Aeronautics and
   Astronautics
555 W. 57th St.
New York, NY 10019
(212) 247-6500

National Aeronautics and Space
   Administration
   Scientific and Technical
   Information Division
300 Seventh St. SW
Washington, DC 20546
(202) 755-1099

PTS F&S Index
   Predicasts, Inc.
11001 Cedar Ave.
Cleveland, OH 44106
(800) 321-6388

Aerospace Industries
   Association of America
1725 DeSales St. NW
Washington, DC 20036
(202) 429-4600

Aerospace Daily
   Murdoch Magazines
1156 15th St. NW
Washington, DC 20005
(202) 822-4600

---

[1]Compiled from *The Encyclopedia of Associations, The Encyclopedia of Business Information Sources,* U.S. government agencies, and other sources.

Aerospace Facts and Figures
  McGraw-Hill, Inc.
  Aerospace and Defense
    Market Focus Group
  1221 Avenue of the Americas
  New York, NY 10020
  (800) 262-4729 or
  (212) 512-2000

Aerospace Education
  Foundation
  1501 Lee Highway
  Arlington, VA 22209
  (703) 247-5839

Aerospace Electrical Society
  Village Station
  P.O. Box 24BB3
  Los Angeles, CA 90024
  (714) 778-1840

Aerospace Industries
  Association of America
  1250 Eye St. NW
  Washington, DC 20005
  (202) 371-8400

Society for the Advancement of
  Material and Process
  Engineering
  P.O. Box 2459
  Covina, CA 91722
  (818) 331-0616

Apparel

Administrative Board—Dress
  Industry (ABDI)
  450 7th Ave., 7th floor
  New York, NY 10001
  (212) 239-2011

Affiliated Dress Manufacturers
  (ADM)
  1440 Broadway
  New York, NY 10018
  (212) 398-9797

Allied Underwear Association
  (AUA)
  100 E. 42d St.
  New York, NY 10017
  (212) 867-5720

American Fashion Association
  (AFA)
  The Dallas Apparel Mart,
  Suite 5442
  Dallas, TX 75258
  (214) 631-0821

Bureau of Wholesale Sales
  Representatives
  1819 Peachtree Rd. NE,
  Suite 210
  Atlanta, GA 30309
  (404) 351-7355

American Fur Industry (AFI)
  363 7th Ave., 7th floor
  New York, NY 10001
  (212) 564-5133

Apparel Guild (AG)
  Gallery 34, Suite 407
  147 W. 33d St.
  New York, NY 10001
  (212) 279-4580

Chamber of Commerce of the
  Apparel Industry (CCAI)
  570 7th Ave.
  New York, NY 10018
  (212) 354-0907

American Apparel
  Manufacturers Association
  (AAMA)
2500 Wilson Blvd., Suite 301
Arlington, VA 22201
(703) 524-1864

Amalgamated Clothing and
  Textile Workers Union
  (ACTWU)
15 Union Sq. W.
New York, NY 10003
(212) 242-0700

Fiber, Fabric and Apparel
  Coalition for Trade (FFACT)
1801 K St. NW, Suite 900
Washington, DC 20006
(202) 862-0517

Apparel Industry Magazine
  Shore Publishing Co.
180 Allen Rd. NE, Suite 300 N
Atlanta, GA 30328
(404) 252-8831

Textile Highlights
  American Textile
  Manufacturers Institute,
  Inc.
1801 K St. NW, Suite 900
Washington, DC 20006
(202) 862-0500

*Highlights of U.S. Export and
  Import Trade*
Bureau of the Census
U.S. Department of
  Commerce
Government Printing Office
Washington, DC 20402
(202) 783-3238

Monthly Labor Review
Bureau of Labor Statistics
U.S. Department of Labor
Government Printing Office
Washington, DC 20402
(202) 783-3238

Overseas Business Reports
Industry and Trade
  Administration
U.S. Department of
  Commerce
Government Printing Office
Washington, DC 20402
(202) 783-3238

International Association of
  Clothing Designers
240 Madison Ave., 12th floor
New York, NY 10016
(212) 685-6602

National Outerwear and
  Sportswear Association
240 Madison Ave.
New York, NY 10016
(212) 686-3440

Beverages

National Licensed Beverage
  Association (NLBA)
4214 King St., W.
Alexandria, VA 22302
(703) 671-7575

National United Affiliated
  Beverage Association
  (NUABA)
P.O. Box 9308
Philadelphia, PA 19139
(215) 748-5670

National United Licensees
Beverage Association
(NULBA)
7141 Frankstown Ave.
Pittsburgh, PA 15208
(412) 241-9344

Can Manufacturers Institute
(CMI)
1625 Massachusetts Ave. NW
Washington, DC 20036
(202) 463-6745

National Beverage Packing
Association (NBPA)
c/o Jerry Testa
200 Daingerfield Rd.
Alexandria, VA 22314-2800
(703) 684-1080

Carbonated Beverage Institute
(CBI)
1101 16th St. NW
Washington, DC 20036
(202) 463-6745

Master Brewers Association of
the Americas (MBAA)
4513 Vernon Blvd.
Madison, WI 53705
(608) 231-3446

National Beverage Dispensing
Equipment Association
(NBDEA)
2011 I St. NW, 5th floor
Washington, DC 20006
(202) 775-4885

National Juice Products
Association (NJPA)
P.O. Box 1531
215 Madison St.
Tampa, FL 33601
(813) 229-1089

National Soft Drink Association
(NSDA)
1101 16th St. NW
Washington, DC 20036
(202) 463-6732

*Hereld's 5000: The Directory of
Leading U. S. Food,
Confectionery, and Beverage
Manufacturers*
S.I.C. Publishing Co.
Whitney Center, Suite 330
200 Leader Hill Dr.
Hamden, CT 06617
(203) 281-6766

*Industry Norms and Key Business
Ratios*
Dun and Bradstreet Credit
Services
One Diamond Hill Rd.
Murray Hill, NJ 07974
(800) 351-3446 or
(201) 665-5330

Foods Adlibra
Foods Adlibra Publications
9000 Plymouth Ave. N.
Minneapolis, MN 55427
(612) 540-3463

Beverage World
Keller International
Publishing Corp.
150 Great Neck Rd.
Great Neck, NY 11021
(516) 829-9210

Beverages
International Beverage
Publications, Inc.
P.O. Box 7406
Overland Park, KS 66207
(913) 341-0020

Beverage Media
Beverage Media Ltd.
161 Sixth Ave.
New York, NY 10013
(212) 620-0100

Impact Beverage Trends in America
M. Shanken Communications, Inc.
387 Park Ave. S.
New York, NY 10016
(212) 684-4224

National Food Review
Economics, Statistics and Cooperatives Service
U.S. Department of Agriculture
Government Printing Office
Washington, DC 20402
(202) 783-3238

FIND/SVP, Inc.
625 Avenue of the Americas
New York, NY 10011
(800) 346-3787 or
(212) 645-4500

Building Materials

Associated Builders and Contractors (ABC)
729 15th St. NW
Washington, DC 20005
(202) 637-8800

Builders' Hardware Manufacturers Association (BHMA)
355 Lexington Ave., 17th floor
New York, NY 10017
(212) 661-4261

National Association of Home Builders of the U.S. (NAHB)
15th and M St. NW
Washington, DC 20005
(202) 822-0200

Door and Hardware Institute (DHI)
7711 Old Springhouse Rd.
McLean, VA 22102
(703) 556-3990

American Building Contractors Association (ABCA)
11100 Valley Blvd., Suite 120
El Monte, CA 91731
(818) 401-0071

National Building Material Distributors Association (NBMDA)
1417 Lake Cook Rd., Suite 130
Deerfield, IL 60015
(708) 945-7201

National Council of the Housing Industry (NCHI)
15th and M St. NW
Washington, DC 20005
(202) 822-0520

Associated Building Material Distributors of America (ABMDA)
7100 E. Lincoln Dr.,
Suite D-220
Scottsdale, AZ 85253
(602) 998-0696

International Conference of Building Officials (ICBO)
5360 S. Workman Mill Rd.
Whittier, CA 90601
(213) 699-0541

Building Officials and Code Administrators International (BOCA) 4051 W. Flossmoor Rd. Country Club Hills, IL 60477-5795 (708) 799-2300

Building and Construction Trades Department AFL-CIO 815 16th St. NW, Suite 603 Washington, DC 20006 (202) 347-1461

Construction Review—Directory of National Trade Associations, Professional Societies, and Labor Unions of the Construction and Building Products Industries U.S. Department of Commerce Building Materials and Construction Division Room 4043 Washington, DC 20230 (202) 377-0132

North American Wholesale Lumber Association— Distribution Directory North American Wholesale Lumber Association 2340 S. Arlington Heights Rd., No. 680 Arlington Heights, IL 60005 (312) 981-8630

Construction Review Industry and Trade Administration U.S. Department of Commerce Government Printing Office Washington, DC 20402 (202) 783-3238

NAHB National Research Center 400 Prince George's Blvd. Upper Marlboro, MD 20772 (301) 249-4000

Small Homes Council— Building Research Council University of Illinois One E. Saint Mary's Rd. Champaign, IL 61820 (217) 333-1801

Current Industrial Reports Bureau of the Census Washington, DC 20233 (301) 763-4100

Chemicals

National Association of Chemical Distributors (NACD) 1200 17th St. NW, Suite 400 Washington, DC 20036 (202) 296-9200

Institute for Chemical Education (ICE) University of Wisconsin Department of Chemistry 1101 University Ave. Madison, WI 53706 (608) 262-3033

Sales Association of the Chemical Industry (SACI) 287 Lackawanna Ave. P.O. Box 2148 West Paterson, NJ 07424 (201) 256-5547

Chemical Manufacturers Association (CMA) 2501 M St. NW Washington, DC 20037 (202) 887-1100

Chemical Marketing Research
  Association (CMRA)
60 Bay St., Suite 702
Staten Island, NY 10301
(718) 876-8800

Chemical Communications
  Association (CCA)
c/o Fleishman-Hillard, Inc.
40 W. 57th St.
New York, NY 10019
(212) 265-9150

Council for Chemical Research
  (CCR)
1620 L St. NW, Suite 825
Washington, DC 20036
(202) 429-3971

American Chemical Society
  (ACS)
1155 16th St. NW
Washington, DC 20036
(202) 872-4600

Chemical Sources Association
  (CSA)
1620 I St. NW, Suite 925
Washington, DC 20006
(202) 293-5800

Chemical Specialties
  Manufacturers Association
  (CSMA)
1913 Eye St. NW
Washington, DC 20006
(202) 872-8110

Chemical Sources—International
Directories Publishing Co.,
  Inc.
P.O. Box 1824
Clemson, SC 29633
(803) 646-7840

Directory of Chemical Producers—
  United States
SRI International
333 Ravenswood Ave.
Menlo Park, CA 94025
(415) 859-3627

CA Search
Chemical Abstracts Service
P.O. Box 3012
Columbus, OH 43210
(614) 447-3600

Annual Review of the Chemical
  Industry
United Nations
Publishing Division
Two United Nations Plaza
New York, NY 10017
(800) 553-3210 or
(212) 963-8297

Current Industrial Reports:
  Inorganic Chemicals
Bureau of the Census
Washington, DC 20233
(301) 763-1584

Quarterly Financial Report for
  Manufacturing, Mining, and
  Trade Corporations
U.S. Federal Trade
  Commission and
U.S. Securities and Exchange
  Commission
Government Printing Office
Washington, DC 20402
(301) 783-3238

Computers

Computer-Aided
  Manufacturing
  International (CAMI)
1250 E. Copeland Rd., No. 500
Arlington, TX 76011
(817) 860-1654

International Council for
  Computer Communications
  (ICCC)
P.O. Box 9745
Washington, DC 20016
(301) 530-7628

Independent Computer
  Consultants Association
  (ICCA)
933 Gardenview Office
  Parkway
St. Louis, MO 63141
(314) 997-4633

National Computer Dealer
  Forum (NCDF)
c/o National Office Products
  Association
301 N. Fairfax St.
Alexandria, VA 22314
(703) 549-9040

National Computer Graphics
  Association (NCGA)
2722 Merrilee Dr., Suite 200
Fairfax, VA 22031
(703) 698-9600

World Computer Graphics
  Association (WCGA)
2033 M St. NW, Suite 399
Washington, DC 20036
(202) 775-9556

ABCD: The Micro Computer
  Industry Association
1515 E. Woodfield Rd.,
  Suite 860
Schaumbrug, IL 60173-5437
(708) 240-1818

Computer and Communications
  Industry Association
  (CCIA)
666 11th St. NW, Suite 600
Washington, DC 20001
(202) 783-0070

The International
  Microcomputer Information
  Exchange (TIMIX)
P.O. Box 201897
Austin, TX 78720
(512) 250-7151

Association for Computer
  Educators (ACE)
c/o Dr. Ben Bauman
College of Business—IDS
James Madison University
Harrisonburg, VA 22807
(703) 568-6189

Computer and Automated
  Systems Association of
  Society of Manufacturing
  Engineers (CASA/SME)
Box 930
1 SME Dr.
Dearborn, MI 48121
(313) 271-1500

Data Processing Management
  Association (DPMA)
505 Busse Hwy.
Park Ridge, IL 60068
(708) 825-8124

IEEE Computer Society
1730 Massachusetts Ave. NW
Washington, DC 20036
(202) 371-0101

Special Interest Group for
Computers and Society
(SIGCAS)
c/o Association for
Computing Machinery
11 W. 42d St.
New York, NY 10036
(212) 869-7440

Software Industry Section of
ADAPSO
c/o ADAPSO
1616 N. Fort Myer, Suite 1300
Arlington, VA 22209-9998
(703) 522-5055

Software Maintenance
Association (SMA)
P.O. Box 12004, No. 297
Vallejo, CA 94590
(707) 643-4423

Association of Computer
Professionals (ACP)
230 Park Ave., Suite 460
New York, NY 10169
(212) 599-3019

National Computer Association
(NCA)
1485 E. Fremont Circle S.
Littleton, CO 80122
(303) 797-3559

Digital Equipment Computer
Users Society (DECUS)
333 South St.
Shrewsbury, MA 01545
(508) 467-5111

North American Computer
Service Association
(NACSA)
100 Silver Beach, No. 918
Daytona Beach, FL 32118
(904) 255-9040

ADAPSO—The Computer
Software and Services
Industry Association
1300 N. 17th St., Suite 300
Arlington, VA 22209
(703) 522-5055

*Applied Science and Technology
Index*
H.W. Wilson Co.
950 University Ave.
Bronx, NY 10452
(800) 367-6770 or
(212) 588-8400

*Computer Industry Forecasts: The
Source for Market Information
on Computers, Peripherals,
and Software*
Data Analysis Group
P.O. Box 4210
Georgetown, CA 95634
(916) 333-4001

Computer Review
GML Corp.
594 Marrett Rd.
Lexington, MA 02173
(617) 861-0515

Computer Shopper
Ziff Davis Publishing Co.
Computer Publications
Division
One Park Ave.
New York, NY 10017
(212) 503-5100

Computerworld
I.D.G. Communications, Inc.
375 Cochituate Rd.
Framingham, MA 01701
(617) 879-0700

IBM Journal of Research and
  Development
IBM Corp.
Armonk, NY 10504
(914) 765-1900

Computers and Office and
  Accounting Machines
Government Printing Office
Washington, DC 20402
(202) 783-3238

Electronics

American Component Dealers
  Association (ACDA)
111 McPherson Church Rd.
Fayetteville, NC 28303
(919) 868-1111

American Electronics
  Association (AEA)
5201 Great America Parkway
Santa Clara, CA 95054
(408) 987-4200

Association of Electronic
  Distributors (AED)
9363 Wilshire Blvd., Suite 217
Beverly Hills, CA 90210
(213) 278-0543

Association of High Tech
  Distributors (AHTD)
1900 Arch St.
Philadelphia, PA 19103
(215) 564-3484

Electronic Industries
  Association (EIA)
2001 Pennsylvania Ave. NW,
  Suite 1100
Washington, DC 20006-1813
(202) 457-4900

Industry Coalition on
  Technology Transfer
  (ICOTT)
1400 L St. NW, 8th floor
Washington, DC 20005-3502
(202) 371-5994

Electronics Representatives
  Association (ERA)
20 E. Huron
Chicago, IL 60611
(312) 649-1333

Semiconductor Equipment and
  Materials International
  (SEMI)
805 E. Middlefield Rd.
Mountain View, CA 94043
(415) 964-5111

Semiconductor Industry
  Association (SIA)
4300 Stevens Creek Blvd.,
  No. 271
San Jose, CA 95129
(408) 246-2711

Surface Mount Equipment
  Manufacturers Association
  (SMEMA)
4113 Barberry Dr.
Lafayette Hill, PA 19444
(215) 825-1008

Association of Old Crows
  (AOC)
1000 N. Payne St.
Alexandria, VA 22314
(703) 549-1600

Audio Engineering Society
  (AES)
60 E. 42d St., Room 2520
New York, NY 10065
(212) 661-8528

IEEE Aerospace and Electronics
  Systems Society (AESS)
c/o Institute of Electrical and
  Electronics Engineers
345 E. 47th St.
New York, NY 10017
(212) 705-7867

IEEE Circuits and Systems
  Society (CSS)
c/o Institute of Electrical and
  Electronics Engineers
345 E. 47th St.
New York, NY 10017
(212) 705-7867

IEEE Consumer Electronics
  Society (CES)
c/o Institute of Electrical and
  Electronics Engineers
345 E. 47th St.
New York, NY 10017
(212) 705-7867

IEEE Industrial Electronics
  Society (IES)
c/o Institute of Electrical and
  Electronics Engineers
345 E. 47th St.
New York, NY 10017
(212) 705-7867

Institute of Electrical and
  Electronics Engineers (IEEE)
345 E. 47th St.
New York, NY 10017
(212) 705-7900

National Engineering
  Consortium (NEC)
303 E. Wacker Dr., Suite 740
Chicago, IL 60601
(312) 938-3500

Aircraft Electronics Association
  (AEA)
P.O. Box 1981
Independence, MO 64055
(816) 373-6565

Survey of Electronics
Leland Schwartz
Merrill Publishing Co.
1300 Alum Creek Dr.
Columbus, OH 43216
(800) 848-1567 or
(614) 258-8441

Electrical and Electronic Abstracts
Institution of Electrical
  Engineers
445 Hoes Lane
Piscataway, NJ 08854

*Electronic Industry Telephone
  Directory*
Harris Publishing Co.
2057-2 Aurora Rd.
Twinsburg, OH 44087
(216) 425-9143

U.S. Electronic Industry Directory
Harris Publishing Co.
2057-2 Aurora Rd.
Twinsburg, OH 44087
(216) 425-9000

INSPEC
Institution of Electrical
  Engineers
Station House
Nightingale Rd.
Hitchin, Herts
England SG5 IRJ

Electronic Business
Cahners Publishing Co., Inc.
275 Washington St.
Newton, MA 02158
(617) 964-3030

Electronic News
Fairchild Publications, Inc.
7 E. 12th St.
New York, NY 10003
(212) 741-4230

Electronic Market Data Book
Electronic Industries
  Association
2001 Pennsylvania Ave. NW
Washington, DC 20006
(202) 457-4900

Food

International Frozen Food
  Association (IFFA)
1764 Old Meadow Lane
McLean, VA 22102
(703) 821-0770

National Frozen Food
  Association (NFFA)
4755 Linglestown Rd.,
  Suite 300
P.O. Box 6069
Harrisburg, PA 17112
(717) 657-8601

Snack Food Association (SFA)
1711 King St., Suite 1
Alexandria, VA 22314
(703) 836-4500

Food Equipment Manufacturers
  Association (FEMA)
401 N. Michigan Ave.
Chicago, IL 60601
(312) 644-6610

Food Service Equipment
  Distribution Association
  (FEDA)
332 S. Michigan Ave.
Chicago, IL 60604
(312) 427-9605

International Food Service
  Distributors Association
  (IFDA)
201 Park Washington Ct.
Falls Church, VA 22046
(703) 532-9400

Food Service Marketing
  Institute (FSMI)
P.O. Box 1265
Lake Placid, NY 12946
(518) 523-2942

International Dairy-Deli
  Association (IDDA)
P.O. Box 5528
313 Price Pl., Suite 202
Madison, WI 53705
(608) 238-7908

International Federation of
  Grocery Manufacturers
  Association (IFGMA)
c/o Grocery Manufacturers of
  America
1010 Wisconsin Ave. NW,
  Suite 800
Washington, DC 20007
(202) 337-9400

American Institute of Food
  Distribution (AIFD)
28-12 Broadway
Fair Lawn, NJ 07410
(201) 791-5570

Association of Food Industries
(AFI)
Sravine Dr.
P.O. Box 776
Matawan, NJ 07747
(201) 583-8188

Educational Foundation of the
National Restaurant
Association (EFNRA)
250 S. Wacker Dr., No. 1400
Chicago, IL 60606
(312) 715-1010

National Soft Serve and Fast
Food Association (NSSFFA)
516 S. Front St.
Chesaning, MI 48616
(517) 845-3336

Food Marketing Institute (FMI)
1750 K St. NW, Suite 700
Washington, DC 20006
(202) 452-8444

National Food Distributors
Association (NFDA)
401 N. Michigan Ave.,
Suite 2400
Chicago, IL 60611
(312) 644-6610

Food and Drug Law Institute
(FDLI)
1000 Vermont Ave., 12th floor
Washington, DC 20005
(202) 371-1420

National Food and Energy
Council (NFEC)
409 Vandiver W., Suite 202
Columbia, MO 65202
(314) 875-7155

Food Research and Action
Center (FRAC)
1319 F St. NW, Suite 500
Washington, DC 20004
(202) 393-5060

Food and Nutrition Board
(FNB)
Institute of Medicine
2101 Constitution Ave. NW
Washington, DC 20418
(202) 334-2238

Food and Allied Service Trades
Department (of AFL-CIO)
(FAST)
815 16th St. NW, Suite 408
Washington, DC 20006
(202) 737-7200

*Hereld's 5000: The Directory of
Leading U.S. Food,
Confectionery, and Beverage
Manufacturers*
S.I.C. Publishing Co.
Whitney Center, Suite 330
200 Leader Hill Dr.
Hamden, CT 06617
(203) 281-6766

Nielsen Retail Index
Nielsen Marketing Research
Nielsen Plaza
Northbrook, IL 60062
(312) 498-6300

Food Industry Newsletter
Newsletters, Inc.
7600 Carter Ct.
Bethesda, MD 20817
(703) 631-2322

Food Research Institute
Stanford University
Stanford, CA 94305
(415) 723-3941

Food and Drug Administration
Shepard's/McGraw-Hill
420 N. Cascade Ave.
Colorado Springs, CO 80901
(800) 525-2474 or
(303) 577-7707

Forest Products

American Hardwood Export
Council (AHEC)
1250 Connecticut Ave. NW,
Suite 200
Washington, DC 20036
(202) 463-2723

American Pulpwood
Association (APA)
1025 Vermont Ave. NW,
Suite 1020
Washington, DC 20005
(202) 347-2900

Western Forest Industries
Association (WFIA)
1500 SW Taylor
Portland, OR 97205
(503) 224-5455

Forest Industries Council (FIC)
1250 Connecticut Ave. NW,
Suite 320
Washington, DC 20036
(202) 463-2460

Hardwood Research Council
(HRC)
P.O. Box 34518
Memphis, TN 38184-0518
(901) 377-1824

National Forest Products
Association (NFPA)
1250 Connecticut Ave. NW,
Suite 200
Washington, DC 20036
(202) 463-2700

North American Wholesale
Lumber Association
(NAWLA)
2340 S. Arlington Heights Rd.,
Suite 680
Arlington Heights, IL 60005
(708) 981-8630

Northwest Forestry Association
(NFA)
1500 SW 1st Ave., Suite 770
Portland, OR 97201
(503) 222-9505

Northwestern Lumbermens
Association (NLA)
1405 N. Lical Dr., Suite 130
Golden Valley, MN 55422
(612) 544-6822

Southeastern Lumber
Manufacturers Association
(SLMA)
P.O. Box 1788
Forest Park, GA 30051
(404) 361-1445

Timber Products Manufacturers
(TPM)
951 E. 3d Ave.
Spokane, WA 99202
(509) 535-4646

American Lumber Standards
Committee (ALSC)
P.O. Box 210
Germantown, MD 20875
(301) 972-1700

Hardwood Distributors
Association (HDA)
1279 N. McLean St.
P.O. Box 12802
Memphis, TN 38182
(901) 274-6887

Wood Products Manufacturers
Association (WPMA)
52 Racette Ave.
Gardner, MA 01440
(508) 632-3923

Association of Consulting
Foresters (ACF)
5410 Grosvenor Lane,
Suite 205
Bethesda, MD 20814
(301) 530-6795

Forest Farmers Association
(FFA)
P.O. Box 95385
Atlanta, GA 30347
(404) 325-2954

National Association of State
Foresters (NASF)
c/o Terri Bates
Hall of States
444 N. Capitol St. NW,
Suite 526
Washington, DC 20001
(202) 624-5415

Forest Products Research
Society
2801 Marshall Ct.
Madison, WI 53705
(608) 231-1361

Association of Western Pulp
and Paper Workers
(AWPPW)
P.O. Box 4566
1430 SW Clay
Portland, OR 97208
(503) 228-7486

Forest
Forest Products Research
Society
2801 Marshall Ct.
Madison, WI 53705
(608) 231-1361

Forest Industries
Miller Freeman Publications
500 Howard St.
San Francisco, CA 94105
(415) 397-1881

Forest Products Laboratory
One Gifford Pinchot Dr.
Madison, WI 53705
(608) 264-5600

*Demand and Price Situation for
Forest Products*
U.S. Forest Service
Government Printing Office
Washington, DC 20402
(202) 783-3238

*U.S. Timber Production, Trade,
Consumption, and Price
Statistics*
Forest Service
U.S. Department of
Agriculture
14th St. and Independence
Ave. SW
Washington, DC 20250
(202) 447-3760

American Forest Council
1250 Connecticut Ave. NW,
Suite 320
Washington, DC 20036
(202) 463-2455

Furniture

National Home Furnishings
Association (NHFA)
P.O. Box 2396
High Point, NC 27261
(919) 883-1650

Contract Furnishings Council
(CFC)
1190 Merchandise Mart
Chicago, IL 60654
(312) 321-0563

American Society of Furniture
Designers (ASFD)
P.O. Box 2688
High Point, NC 27261
(919) 884-4074

American Furniture
Manufacturers Association
(AFMA)
P.O. Box HP-7
High Point, NC 27261
(919) 884-5000

International Home Furnishings
Marketing Association
(IHFMA)
P.O. Box 5687
High Point, NC 27262
(919) 889-0203

National Unfinished Furniture
Institute (NUFI)
1850 Oak St.
Northfield, IL 60093
(708) 446-8434

National Wholesale Furniture
Association (NWFA)
P.O. Box 2482
164 S. Main St., Suite 404
High Point, NC 27261
(919) 884-1566

National Office Products
Association (NOPA)
301 N. Fairfax St.
Alexandria, VA 22314
(703) 549-9040

Furniture/Today
Communications Today
Publishing, Ltd.
200 S. Main St.
High Point, NC 27261
(919) 889-0113

Retail Trade, Annual Sales, Year-End Inventories, and Accounts Receivable by Kind of Retail Store
Bureau of the Census
U.S. Department of
Commerce
Washington, DC 20233
(202) 763-4040

Industrial and Farm Equipment

American Boiler Manufacturers
Association (ABMA)
950 N. Glebe Rd., Suite 160
Arlington, VA 22203
(703) 522-7350

American Gear Manufacturers
Association (AGMA)
1500 King St., Suite 201
Alexandria, VA 22314
(703) 684-0211

American Machine Tool
Distributors' Association
(AMTDA)
1335 Rockville Pike
Rockville, MD 20852
(301) 738-1200

American Supply and
Machinery Manufacturers
Association (ASMMA)
Thomas Associates, Inc.
1230 Keith Building
Cleveland, OH 44115
(216) 241-7333

American Textile Machinery
Association (ATMA)
7297 Lee Highway, Suite N
Falls Church, VA 22042
(703) 533-9251

Associated Equipment
Distributors (AED)
615 W. 22d St.
Oak Brook, IL 60521
(708) 574-0650

Fabricators and Manufacturers
Association, International
(FMA)
5411 E. State St.
Rockford, IL 61108
(815) 399-8700

Hydraulic Tool Manufacturers
Association (HTMA)
P.O. Box 1337
Milwaukee, WI 53201
(414) 633-3454

Industrial Distribution
Association (IDA)
3 Corporate Sq., Suite 201
Atlanta, GA 30329
(404) 325-2776

Manufacturers Alliance for
Productivity and
Innovation (MAPI)
1200 18th St. NW, Suite 400
Washington, DC 20036
(202) 331-8430

Surface Mount Technology
Association (SMTA)
5200 Wilson Rd., Suite 100
Edina, MN 55424
(612) 920-SMTA

Industrial Fasteners Institute
(IFI)
1505 E. Ohio Building
Cleveland, OH 44114
(216) 241-1482

Equipment Manufacturers
Institute (EMI)
c/o Tim Metzger
10 S. Riverside Plaza,
Suite 1220
Chicago, IL 60606
(312) 321-1470

Farm Equipment Manufacturers
Association (FEMA)
243 N. Lindbergh Blvd.
St. Louis, MO 63141
(314) 991-0702

North American Equipment
Dealers Association
(NAEDA)
10877 Watson Rd.
St. Louis, MO 63127
(314) 821-7220

*Agricultural Engineering
Abstracts*
CAB International North
America
845 N. Park Ave.
Tucson, AZ 85719
(800) 528-4841 or
(602) 621-7897

*Farm Equipment Wholesalers
Association Directory*
Farm Equipment Wholesalers
Association
1927 Keokuk St.
Iowa City, IA 52240
(319) 354-5156

Agricola
    U.S. National Agricultural
    Library
    Beltsville, MD 20705
    (301) 344-3813

NTIS Bibliographic Database
    U.S. Department of
    Commerce
    National Technical
    Information Service
    5285 Port Royal Rd.
    Springfield, VA 22161
    (703) 487-4630

Farm Equipment
    Johnson Hill Press, Inc.
    1233 Janesville Ave.
    Fort Atkinson, WI 53538
    (414) 563-6388

Annual Survey of Manufacturers
    Bureau of the Census
    U.S. Department of
    Commerce
    Government Printing Office
    Washington, DC 20402
    (202) 783-3238

Farm Equipment Wholesalers
    Association
    P.O. Box 1347
    Iowa City, IA 52240
    (319) 354-5156

Industrial Equipment News
    Thomas Publishing Co.
    250 W. 34th St.
    New York, NY 10119
    (212) 868-5661

Metals and Metal Products

Aluminum Association (AA)
    900 19th St. NW, Suite 300
    Washington, DC 20006
    (202) 862-5100

American Copper Council
    (ACC)
    333 Rector Pl., Suite 10P
    New York, NY 10280
    (212) 945-4990

American Institute for Imported
    Steel (AIIS)
    11 W. 42d St., Suite 3002
    New York, NY 10036-8002
    (212) 921-1765

Association of Steel Distributors
    (ASD)
    401 N. Michigan Ave.
    Chicago, IL 60611-4390
    (312) 644-6610

Metal Building Manufacturers
    Association (MBMA)
    c/o Charles M. Stockinger
    Thomas Associates, Inc.
    1230 Keith Building
    Cleveland, OH 44115
    (216) 241-7333

Specialty Steel Industry of the
    United States (SSIUS)
    3050 K St. NW, 4th floor
    Washington, DC 20007
    (202) 342-8400

Steel Manufacturers Association
    (SMA)
    815 Connecticut Ave. NW,
    no. 304
    Washington, DC 20006
    (202) 342-1160

Metal Finishing Suppliers'
    Association (MFSA)
    801 N. Cass Ave., Suite 300
    Westmount, IL 60555
    (708) 887-0797

American Bureau of Metal
   Statistics (ABMS)
P.O. Box 1405
400 Plaza Dr.
Secaucus, NJ 07094
(201) 863-6900

Association of Iron and Steel
   Engineers (AISE)
3 Gateway Center, Suite 2350
Pittsburgh, PA 15222
(412) 281-6323

ASM International
9639 Kinsman
Materials Park, OH 44073
(216) 338-5151

Industry Council for Tangible
   Assets (ICTA)
25 E St. NW, 8th floor
Washington, DC 20001
(202) 783-3500

Metal Fabricating Institute
   (MFI)
P.O. Box 1178
Rockford, IL 61105
(815) 965-4031

Metal Trades Department (of
   AFL-CIO) (MTD)
503 AFL-CIO Building
815 16th St. NW
Washington, DC 20006
(202) 347-7255

American Metal Market
7 E. 12th St.
New York, NY 10003
(212) 741-4140

Journal of Metals
Metallurgical Society, Inc.
420 Commonwealth Dr.
Warrendale, PA 15086
(412) 776-9070

Metals Week
McGraw-Hill Publishing Co.
1221 Avenue of the Americas
New York, NY 10020
(800) 262-4769 or
(212) 512-2000

Metal Statistics
Fairchild Publications, Inc.
7 E. 12th St.
New York, NY 10003
(212) 741-4140

Mining, Crude Oil Production

Colorado Mining Association
   (CMA)
1600 Broadway, Suite 1340
Denver, CO 80202
(303) 894-0536

Northwest Mining Association
   (NWMA)
414 Peyton Building
Spokane, WA 99201
(509) 624-1158

Society for Mining, Metallurgy,
   and Exploration (SME, Inc.)
P.O. Box 625005
Littleton, CO 80162
(303) 973-9550

Center for Alternative Mining
   Development Policy
   (CAMDP)
210 Avon St., Suite 9
La Crosse, WI 54603
(608) 784-4399

Pittsburgh Coal Mining
   Institute of America
   (PCMIA)
4800 Forbes Ave.
Pittsburgh, PA 15213
(412) 621-4500

Rocky Mountain Coal Mining
Institute (RMCMI)
3000 Youngfield, no. 324
Lakewood, CO 80215-6545
(303) 238-9099

Pennsylvania Grade Crude Oil
Association (PGCOA)
c/o Pringle Powder Co.
Box 201
Bradford, PA 16701
(814) 368-8172

American Society for Surface
Mining and Reclamation
(ASSMR)
21 Grandview Dr.
Princeton, WV 24740
(304) 425-8332

American Institute of Mining,
Metallurgical and
Petroleum Engineers
(AIME)
345 E. 47th St., 14th floor
New York, NY 10017
(212) 705-7695

Mining and Metallurgical
Society of America (MMSA)
210 Post St., Suite 1102
San Francisco, CA 94108
(415) 398-6925

*American Mining Congress
Journal*
American Mining Congress
1920 N St. NW
Washington, DC 20036
(202) 861-2800

*Colorado School of Mines
Quarterly*
Colorado School of Mines
Press
Golden, CO 80401
(303) 273-3600

Earth and Mineral Sciences
Penn State College of Earth
and Mineral Sciences
Pennsylvania State University
116 Deike Building
University Park, PA 16802
(814) 863-4667

Census of Mineral Industries
Bureau of the Census
Washington, DC 20233
(301) 763-4100

Minerals and Materials
U.S. Bureau of the Census
Washington, DC 20233
(202) 634-1001

Mining Machinery and Equipment
U.S. Bureau of the Census
Washington, DC 20233
(301) 763-4100

U.S. Bureau of Mines
2401 E St. NW
Washington, DC 20241
(202) 634-1001

Motor Vehicles and Parts

American Automobile
Association (AAA)
1000 AAA Dr.
Heathrow, FL 32746-5063
(407) 444-7000

American International
Automobile Dealers
Association (AIADA)
1128 16th St. NW
Washington, DC 20036
(202) 659-2561

National Automobile Dealers
    Association (NADA)
8400 Westpark Dr.
McClean, VA 22102
(703) 827-7407

Motor Vehicle Manufacturers
    Association of the United
    States (MVMA)
7430 2d Ave., Suite 300
Detroit, MI 48202
(313) 872-4311

Automotive Body Parts
    Association (ABPA)
2500 Wilcrest Dr., Suite 510
Houston, TX 77042-2752
(713) 977-5551

Automotive Cooling Systems
    Institute (ACSI)
300 Sylvan Ave.
P.O. Box 1638
Englewood Cliffs, NJ 07632
(201) 569-8500

Automotive Engine Rebuilders
    Association (AERA)
330 Lexington Dr.
Buffalo Grove, IL 60089-6998
(708) 541-6550

Automotive Exhaust Systems
    Manufacturers Council
    (AESMC)
300 Sylvan Ave.
P.O. Box 1638
Englewood Cliffs, NJ 07632
(201) 569-8500

Automotive Parts and
    Accessories Association
    (APAA)
5100 Forbes Blvd.
Lanham, MD 20706
(301) 459-9110

Automotive Refrigeration
    Products Institute (ARPI)
4600 E. West Highway,
    Suite 300
Bethesda, MD 20814
(301) 657-2774

Brake System Parts
    Manufacturers Council
    (BSPMC)
300 Sylvan Ave.
P.O. Box 1638
Englewood Cliffs, NJ 07632
(201) 569-8500

Filter Manufacturers Council
    (FMC)
300 Sylvan Ave.
P.O. Box 1638
Englewood Cliffs, NJ 07632
(201) 569-8500

National Automotive Parts
    Association (NAPA)
2999 Circle 75 Parkway
Atlanta, GA 30339
(404) 956-2200

Automotive Market Research
    Council (AMRC)
300 Sylvan Ave.
P.O. Box 1638
Englewood Cliffs, NJ 07632
(607) 257-6700

Championship Association of
    Mechanics (CAM)
P.O. Box 7
Howard, CO 81233
(719) 942-3611

Automobile Quarterly
Automobile Quarterly, Inc.
420 N. Park Rd.
Wyomissing, PA 19610
(215) 325-8444

*MVMA Motorvehicle Facts and
  Figures*
Motor Vehicle Manufacturers
  Association of the U.S., Inc.
7430 Second Ave., Suite 300
Detroit, MI 48202
(313) 872-4311

*Dictionary of Automotive
  Technology*
VCH Publishers, Inc.
220 E. 23d St.
New York, NY 10010
(800) 422-8824 or
(212) 683-8333

Automotive Executive
National Automobile Dealers
  Association
8400 Westpark Dr.
McLean, VA 22102
(703) 821-7150

Motor News Analysis
News Analysis, Inc.
32068 Olde Franklin Dr.
Farmington Hills, MI 48018
(313) 851-1377

Automotive Market Report
Automotive Auction
  Publishing, Inc.
1101 Fulton Building
Pittsburgh, PA 15222
(412) 281-2338

Annual Survey of Manufacturers
Bureau of the Census
Washington, DC 20233
(301) 763-4100

*Automotive Industries Statistical
  Issue*
Chilton Co.
Chilton Way
Radnor, PA 19089
(800) 345-1214 or
(215) 964-4000

Automotive Aftermarket News
Stanley Publishing Co.
200 Madison Ave., Suite 2104
Chicago, IL 60601
(312) 332-0210

Motor and Equipment
  Manufacturers Association
300 Sylvan Ave.
Englewood Cliffs, NJ 07632
(201) 569-8500

National Automotive Parts
  Association
2999 Circle 75 Parkway
Atlanta, GA 30339
(404) 956-2200

Petroleum Refining

American Independent Refiners
  Association (AIRA)
649 S. Olive St., Suite 500
Los Angeles, CA 90014
(213) 624-8407

Association of Petroleum
  Re-refiners (APR)
P.O. Box 427
Buffalo, NY 14205
(716) 855-2212

National Petroleum Refiners
  Association (NAPRA)
1899 L St. NW, Suite 1000
Washington, DC 20036
(202) 457-0480

National Petroleum News
Hunter Publishing Co.
950 Lee St.
Des Plains, IL 60016
(708) 296-0770

Basic Petroleum Data Book
American Petroleum Institute
275 7th Ave.
New York, NY 10001
(212) 366-4040

*The Oil and Gas Producing
Industry in Your State*
Independent Petroleum
Association of America
Petroleum Independent
Publishers, Inc.
1101 16th St. NW
Washington, DC 20036
(202) 857-4766

Oil/Energy Statistics Bulletin
Oil Statistics Co.
P.O. Box 127
Babson Park, MA 02157
(617) 651-8126

*Reserves of Crude Oil, Natural
Gas Liquids, and Natural Gas
in the United States and
Canada and U.S. Productive
Capacity*
American Gas Association
1515 Wilson Blvd.
Arlington, VA 22209
(703) 841-8400

Oil and Gas Reporter
Matthew Bender & Co., Inc.
11 Penn Plaza
New York, NY 10017
(800) 223-1940 or
(212) 967-7707

Pharmaceuticals

American Pharmaceutical
Association (APhA)
2215 Constitution Ave. NW
Washington, DC 20037
(202) 628-4410

National Council of State
Pharmaceutical Association
Executives (NCSPAE)
c/o Paul Gacanti
Virginia Pharmaceutical
Assn.
3119 W. Clay St.
Richmond, VA 23230
(804) 355-7941

National Pharmaceutical
Association (NPhA)
Howard University
College of Pharmacy and
Pharmacal Sciences
Washington, DC 20059
(202) 328-9229

American Council on
Pharmaceutical Education
(ACPE)
311 W. Superior St., Suite 512
Chicago, IL 60610
(312) 664-3575

Academy of Pharmaceutical
Research and Science
(APRS)
c/o Naomi U. Kaminsky
American Pharmaceutical
Association
2215 Constitution Ave. NW
Washington, DC 20037
(202) 628-4410

National Pharmaceutical
Council (NPC)
1894 Preston White Dr.
Reston, VA 22041
(703) 620-6390

Generic Pharmaceutical
Industry Association
(GPIA)
200 Madison Ave., Suite 2404
New York, NY 10016
(212) 683-1881

National Association of
Pharmaceutical
Manufacturers (NAPM)
747 3d Ave.
New York, NY 10017
(212) 838-3720

Pharmaceutical Manufacturers
Association (PMA)
1100 15th St. NW
Washington, DC 20005
(202) 835-3400

National Wholesale Druggists'
Association (NWDA)
105 Oronoco St.
P.O. Box 238
Alexandria, VA 22314
(703) 684-6400

*Pharmaceutical Marketers
Directory*
CPS Communications, Inc.
7200 W. Camino Real,
Suite 215
Boca Raton, FL 33433
(305) 368-9301

Physicians' Desk Reference
Medical Economics Co., Inc.
680 Kinderkamack Rd.
Oradell, NJ 07649
(800) 223-0581 or
(201) 262-3030

Food and Drug Administration
Shepard's/McGraw-Hill
420 N. Cascade Ave.
Colorado Springs, CO 80901
(800) 525-2474 or
(303) 577-7707

Publishing/Printing

American Book Producers
Association (ABPA)
41 Union Square W., room 936
New York, NY 10003
(212) 645-2368

American Business Press (ABP)
675 3d Ave., Suite 400
New York, NY 10017
(212) 661-6360

American Newspaper
Publishers Association
(ANPA)
The Newspaper Center
Box 17407
Dulles International Airport
Washington, DC 20041
(703) 648-1000

International Newspaper
Marketing Association
(INMA)
P.O. Box 17422
Dulles International Airport
Washington, DC 20041
(703) 648-1094

Magazine Publishers of
America (MPA)
575 Lexington Ave.
New York, NY 10022
(212) 752-0055

National Business Circulation
Association (NBCA)
Act III Publishing
c/o Steve Wigginton
401 Park Ave. S
New York, NY 10016
(212) 545-5140

National Newspaper Publishers
Association (NNPA)
948 National Press Building,
room 948
Washington, DC 20045
(202) 662-7324

Periodical and Book Association
of America (PBAA)
120 E. 34th St., Suite 7-K
New York, NY 10016
(712) 689-4952

Periodicals Institute
P.O. Box 899
West Caldwell, NJ 07007
(201) 882-1130

Label Printing Industries of
America (LPIA)
100 Daingerfield Rd.
Alexandria, VA 22314
(703) 519-8122

International Financial Printers
Association (IFPA)
100 Daingerfield Rd.
Alexandria, VA 22314
(703) 519-8122

National Association of Printers
and Lithographers (NAPL)
780 Palisade Ave.
Teaneck, NJ 07666
(201) 342-0700

National State Printing
Association (NSPA)
c/o Council of State
Governments
Iron Works Pike
P.O. Box 11910
Lexington, KY 40578
(606) 231-1874

Screen Printing Association
International (SPAI)
10015 Main St.
Fairfax, VA 22031
(703) 385-1335

Printing, Publishing, and Media
Workers Sector
Communications Workers of
America (CWA)
1925 K St. NW, Suite 400
Washington, DC 20006
(202) 728-2326

Machine Printers and Engravers
Association of the United
States (MPEA)
690 Warren Ave.
East Providence, RI 02914
(401) 438-5849

American Printer
Maclean-Hunter Publishing
Co.
29 N Wacker Dr.
Chicago, IL 60606
(800) 621-9907 or
(312) 726-7907

*Publishers Directory: A Guide to
New and Established, Private
and Special-Interest, Avant
Garde and Alternative,
Organization and Association,
Government and Institution
Presses*
Gale Research Inc.
835 Penobscot Bldg.
Detroit, MI 48226-4094
(800) 877-GALE or
(313) 962-2242

Publisher's Weekly
Bowker Magazine Group
Cahners Magazine Division
249 W. 17th St.
New York, NY 10011
(800) 669-1002 or
(212) 645-9700

Rubber and Plastic Products

Rubber Manufacturers
Association
1400 K St. NW
Washington, DC 20005
(202) 682-4800

International Institute of
Synthetic Rubber Producers
(IISRP)
2077 S. Gessner Rd., Suite 133
Houston, TX 77063-1123
(713) 783-7511

Rubber Trade Association of
New York (RTA)
17 Battery Pl.
New York, NY 10004
(212) 344-7776

American Laminators
Association (ALA)
419 Norton Building
Seattle, WA 98104
(206) 622-0666

American Society of
Electroplated Plastics
(ASEP)
1101 14th St. NW, Suite 1100
Washington, DC 20005
(202) 371-1323

Chemical Fabrics and Film
Association (CFFA)
c/o Thomas Associates, Inc.
1230 Keith Building
Cleveland, OH 44115
(216) 241-7323

National Association of Plastics
Distributors (NAPD)
6333 Long St., Suite 340
Shawnee, KS 66216
(913) 268-6273

Polyurethane Division
Society of the Plastics
Industry (PDSPI)
355 Lexington Ave.
New York, NY 10017
(212) 351-5425

Plastic and Metal Products
Manufacturers Association
(PMPMA)
225 W. 34th St., Suite 2002
New York, NY 10001
(212) 564-2500

Polyurethane Manufacturers
Association (PMA)
Building C, Suite 20
800 Roosevelt Rd.
Glen Ellyn, IL 60137
(708) 858-2670

Plastics Recycling Foundation
(PRF)
1275 K St. NW, Suite 500
Washington, DC 20005
(202) 371-5200

Plastics Education Foundation
(PEF)
c/o Society of Plastics
Engineers
14 Fairfield Dr.
Brookfield, CT 06804
(203) 775-0471

Society of Plastics Engineers
(SPE)
14 Fairfield Dr.
Brookfield, CT 06804
(203) 775-0471

Plastics Institute of America
(PIA)
Stevens Institute of
Technology
Castle Point Station
Hoboken, NJ 07030
(201) 420-5553

*Rubber and Plastics News—
Rubbicana Issue*
Crain Communications, Inc.
740 N. Rush St.
Chicago, IL 60611
(312) 649-5200

*Rubbicana: Directory of North
American Rubber Product
Manufacturers and Rubber
Industry Suppliers*
Crain Communications, Inc.
1725 Merriman Rd., Suite 300
Akron, OH 44313
(216) 836-9180

Dictionary of Rubber
Elsevier Science Publishing
Co.
655 Avenue of the Americas
New York, NY 10010
(212) 989-5800

Rubber World
1867 W. Market St.
Akron, OH 44313
(216) 864-2122

*Rubber: Production, Shipments,
and Stocks*
U.S. Bureau of the Census
Washington, DC 20233
(301) 763-4100

*Plastics Technology
Manufacturing Handbook and
Buyers' Guide*
Bill Communications, Inc.
633 Third Ave.
New York, NY 10017
(800) 343-1732 or
(212) 986-4800

Plastics World—Plastics Directory
Cahners Publishing Co., Inc.
1350 E. Touhy Ave.
Des Plaines, IL 60018
(312) 635-8800

Dictionary of Plastics Technology
VCH Publishers Inc.
220 E. 23d St.
New York, NY 10010
(800) 422-8824 or
(212) 683-8333

Modern Plastics
McGraw-Hill, Inc.
1221 Avenue of the Americas
New York, NY 10020
(800) 262-4729 or
(212) 512-2000

*Plastics News: Crain's
International Newspaper for
the Plastics Industry*
Crain Communications Inc.
1725 Merriman Rd., Suite 300
Akron, OH 44313
(216) 836-9180

Plastics World
Cahners Publishing Co., Inc.
275 Washington St.
Newton, MA 02158
(617) 964-3030

Plastics Institute of America
Stevens Institute of
Technology
Castle Point Station
Hoboken, NJ 07030
(201) 420-5553

Scientific and
Photographic Equipment

Commission on Professionals in
Science and Technology
(CPST)
1500 Massachusetts Ave. NW,
Suite 831
Washington, DC 20005
(202) 223-6995

National Association for
Science, Technology and
Society (NASTS)
117 Willard Building
University Park, PA 16802
(814) 865-9951

Association for Science,
Technology and Innovation
(ASTI)
P.O. Box 1242
Arlington, VA 22210
(703) 451-6948

National Association of Photo
Equipment Technicians
(NAPET)
3000 Picture Pl.
Jackson, MI 49201
(517) 788-8100

Carnegie Commission on
Science, Technology, and
Government (CCSTG)
10 Waverly Pl.
New York, NY 10003
(212) 998-2150

United Nations Centre for
Science and Technology for
Development (CSTD)
1 United Nations Plaza, DCI-
10th floor
New York, NY 10017
(212) 963-8435

Guide to Scientific Instruments
American Association for the
Advancement of Science
1333 H St. NW, 8th floor
Washington, DC 20005
(202) 326-6446

*Standards and Practices for
Instrumentation*
Instrument Society of
America
67 Alexander Dr.
Research Triangle Park, NC
27709
(919) 549-8411

Scisearch
Institute for Scientific
Information
3501 Market St.
Philadelphia, PA 19104
(800) 523-1850 or
(215) 386-0100

Review of Scientific Instruments
American Institute of Physics
335 E. 45th St.
New York, NY 10017
(212) 661-9404

*Selected Instruments and Related
Products*
U.S. Bureau of the Census
Washington, DC 20233
(301) 763-4100

Instrument Society of America
67 Alexander Dr.
Research Triangle Park, NC
27709
(919) 549-8411

Scientific Apparatus Makers
Association
1101 16th St. NW
Washington, DC 20036
(202) 223-1360

Journal of Imaging Technology
SPSE: The Society for Imaging
Science and Technology
7003 Kilworth Lane
Springfield, VA 22151
(703) 642-9090

Photographic Society of
America
3000 United Founders Blvd.,
Suite 103
Oklahoma City, OK 73112
(405) 843-1437

Soaps, Cosmetics

Cosmetic Industry Buyers and
Suppliers (CIBS)
c/o Joseph A. Palazzolo
36 Lakeville Rd.
New Hyde Park, NY 11040
(516) 775-0220

Independent Cosmetic
Manufacturers and
Distributors (ICMAD)
1230 W. Northwest Hwy.
Palatine, IL 60067
(708) 991-4499

Cosmetic, Toiletry and
Fragrance Association
(CTFA)
1110 Vermont Ave. NW,
Suite 800
Washington, DC 20005
(202) 331—1770

Soap and Detergent Association
(SDA)
475 Park Ave. S.
New York, NY 10016
(212) 725-1262

Cosmetic World News
130 Wigmore St.
London, England W1H 0AT

Cosmetics and Toiletries
Allured Publishing Corp.
214 W. Willow
Wheaton, IL 60189
(312) 653-2155

Drug and Cosmetic Industry
Edgell Communications, Inc.
7500 Old Oak Blvd.
Cleveland, OH 44130
(216) 243-8100

*U.S. Toiletries and Cosmetics
Industry*
Off-the-Shelf Publications,
Inc.
2171 Jericho Turnpike
Commack, NY 11725
(516) 462-2410

*Soap/Cosmetics/Chemical
Specialties*
MacNair Publications, Inc.
101 W. 31st Ave.
New York, NY 10001
(212) 279-4455

Telecommunications

North American Telecommunications Association— Telecommunications Sourcebook
North American Telecommunications Association
2000 M St., NW, Suite 550
Washington, DC 20036
(800) 538-6286 or
(202) 296-9800

Telecommunications Systems and Services Directory
Gale Research Inc.
835 Penobscott Bldg.
Detroit, MI 48226-4094
(800) 877-GALE or
(313) 962-2242

Computer and Telecommunications Acronyms
Gale Research Inc.
836 Penobscot Bldg.
Detroit, MI 48226-4094
(800) 877-GALE or
(313) 962-2242

AT&T Technical Journal
AT&T Bell Laboratories
101 JFK Parkway
Short Hills, NJ 07078
(201) 564-4280

Telecommunications
Horizon-House-Microwave, Inc.
685 Canton St.
Norwood, MA 02060
(617) 769-9750

Telecommunications Reports
Business Research Publications, Inc.
1036 National Press Bldg.
Washington, DC 20045
(202) 347-2654

Telecommunications Week
Business Research Publications, Inc.
1036 National Press Bldg.
Washington, DC 20045
(202) 347-2654

Communications Equipment, Including Telephone, Telegraph, and Other Electronic Systems and Equipment
Government Printing Office
Washington, DC 20402
(202) 783-3238

Competitive Telecommunications Assn.
120 Maryland Ave. NE
Washington, DC 20002
(202) 546-9022

Datapro Reports on Telecommunications
Datapro Research Corp.
1805 Underwood Blvd.
Delran, NJ 08075
(609) 764-0100

Directory of Top Computer Executives
Applied Computer Research, Inc.
P.O. Box 9280
Phoenix, AZ 85068
(800) 234-2227 or
(602) 995-5929

Moody's Public Utility Manual
Moody's Investors Service,
Inc.
99 Church St.
New York, NY 10007
(212) 553-0300

Communications News
Communication News
111 E. Wacker Dr., 16th floor
Chicago, IL 60601
(312) 938-2300

Communications Week
CMP Publications, Inc.
600 Community Dr.
Manhasset, NY 11030
(516) 365-4600

Telephone Management Strategist
Buyers Laboratory, Inc.
20 Railroad Ave.
Hackensack, NJ 07601
(201) 488-0404

Telephony
Telephony Publishing Corp.
55 E. Jackson Blvd.
Chicago, IL 60604
(312) 922-2435

*Annual Statistical Reports of
Independent Telephone
Companies*
Federal Communications
Commission
1919 M St. NW
Washington, DC 20554
(202) 632-7000

Independent Telephone Statistics
United States Telephone
Association
900 19th St. NW, Suite 800
Washington, DC 20006
(202) 835-3100

*Quarterly Operating Data of 68
Telephone Carriers*
Federal Communications
Commission
1919 M St. NW
Washington, DC 20554
(202) 632-7000

*Standard & Poor's Industry
Surveys*
Standard & Poor's Corp.
25 Broadway
New York, NY 10004
(212) 208-8714

U.S. Industrial Outlook
Industry and Trade
Administration
U.S. Department of
Commerce
Government Printing Office
Washington, DC 20401
(202) 783-3238

National Telephone
Cooperative Association
2626 Pennsylvania Ave. NW
Washington, DC 20037
(202) 298-2300

United States Telephone
Association
900 19th St. NW, Suite 800
Washington, DC 20006
(202) 835-3100

Textiles

American Fiber, Textile,
Apparel Coalition (AFTAC)
1801 K St. NW, Suite 900
Washington, DC 20006
(202) 862-0500

American Reuseable Textile
  Association (ARTA)
P.O. Box 1073
Largo, FL 34294
(314) 889-1360

Knitted Textile Association
  (KTA)
386 Park Ave. S.
New York, NY 10016
(212) 689-3807

Textile Association of Los
  Angeles (TALA)
110 E. 9th, room C-765
Los Angeles, CA 90079
(213) 627-6173

Northern Textile Association
  (NTA)
230 Congress St.
Boston, MA 02110
(617) 542-8220

Southern Textile Association
  (STA)
509 Francisca Lane
P.O. Box 190
Cary, NC 27512
(919) 467-1655

Textile Distributors Association
  (TDA)
45 W. 36th St., 3d floor
New York, NY 10018
(212) 563-0400

Textile Information Users
  Council (TIUC)
c/o Trudy Craven
Milliken Research Corp.
P.O. Box 5521
Spartanburg, SC 29304
(803) 573-1589

Textile Quality Control
  Association (TQCA)
P.O. Box 76501
Atlanta, GA 30328
(404) 252-9037

Textile Fibers and By-Products
  Association (TFBPA)
P.O. Box 11065
Charlotte, NC 28220
(704) 527-5593

Textile Research Institute (TRI)
P.O. Box 625
Princeton, NJ 08542
(609) 924-3150

American Textile Manufacturers
  Institute (ATMI)
1801 K St. NW, Suite 900
Washington, DC 20006
(202) 862-0500

Institutional and Service Textile
  Distributors Association
  (ISTDA)
93 Standish Rd.
Hillsdale, NJ 07642
(201) 664-4600

American Fiber Manufacturers
  Association (AFMA)
1150 17th St. NW
Washington, DC 20036
(202) 296-6508

National Textile Processors
  Guild (NTPG)
75 Livingston St.
Brooklyn, NY 11201
(718) 875-2300

Institute of Textile Technology
  (ITT)
P.O. Box 391
Charlottesville, VA 22902
(804) 296-5511

Textile Converters Association
  (TCA)
100 E. 42d St.
New York, NY 10017
(212) 867-5720

United Textile Workers of
  America (UTWA)
P.O. Box 749
Voorhees, NJ 08043
(609) 772-9699

World Textile Abstracts
Shirley Institute
Charlton St.
Manchester, England M1 3FH

America's Textiles International-
  Directory
Billian Publishing Co.
2100 Powers Ferry Rd.
Atlanta, GA 30339
(404) 955-5656

Fairchild's Textile and Apparel
  Financial Directory
Fairchild Publications, Inc.
7 E. 12th St.
New York, NY 10003
(212) 741-4280

Textile Research Journal
Textile Research Institute
601 Prospect Ave.
Princeton, NJ 08542
(609) 924-3150

Textile World
Maclean Publishing Co.
  Textile Publications
4170 Ashford-Dunwood Rd.,
  Suite 420
Atlanta, GA 30319
(404) 252-0626

Census of Manufacturers
U.S. Bureau of the Census
Government Printing Office
Washington, DC 20402
(202) 783-3238

Textile Distributors Association
45 W. 36th St., 3d floor
New York, NY 10018
(212) 563-0400

Tobacco

Tobacco Associates (TA)
1725 K St. NW, Suite 512
Washington, DC 20006
(202) 828-9144

Burley and Dark Leaf Tobacco
  Association (BDLTE)
1100 17th St. NW
Washington, DC 20036
(202) 296-6820

Tobacco Association of United
  States (TAUS)
3716 National Dr., Suite 114
Raleigh, NC 27612
(919) 782-5151

Cigar Association of America
  (CAA)
1100 17th St. NW, Suite 504
Washington, DC 20036
(202) 223-8204

Tobacconists' Association of
  America (TAA)
c/o Milan Brothers
106 S. Jefferson St.
Roanoke, VA 24011
(703) 344-5191

Council for Tobacco Research—
USA (CTR-USA)
900 Third Ave.
New York, NY 10022
(212) 421-8885

Leaf Tobacco Exporters
Association (LTEA)
3716 National Dr., Suite 114
Raleigh, NC 27612
(919) 782-5151

Tobacco Merchants Association
of United States (TMA)
P.O. Box 8019
231 Clarksville Rd.
Princeton, NJ 08543
(609) 275-4900

Tobacco Institute (TI)
1875 I St. NW, Suite 800
Washington, DC 20006
(202) 457-4800

National Association of Tobacco
Distributors (NATD)
1199 N. Fairfax St., Suite 701
Alexandria, VA 22314
(703) 683-8336

Tobacco Growers' Information
Committee (TGIC)
P.O. Box 12300
Raleigh, NC 27605
(919) 821-0390

National Cigar Leaf Tobacco
Association (NCLTA)
1100 17th St. NW
Washington, DC 20036
(202) 296-6820

Tobacco Industry
Labor/Management
Committee (TILMC)
c/o Dalton Joe Masterson
10401 Connecticut Ave.
Kensington, MD 20895
(301) 933-8600

*Tobacco Barometer: Cigarettes,
Cigars*
Tobacco Merchants
Association of the United
States
P.O. Box 8019
Princeton, NJ 08543
(609) 275-4900

Tobacco Reporter
Specialized Agricultural
Publications, Inc.
3000 Highwoods Blvd.,
Suite 300
Raleigh, NC 27625
(919) 872-5040

Tobacco and Health Research
Institute
University of Kentucky
Cooper and Alumni Dr.
Lexington, KY 40546
(606) 257-4600

Tobacco Retailers Almanac
Retail Tobacco Dealers of
America
107 E. Baltimore St.
Baltimore, MD 21202
(301) 547-6996

Transportation Equipment

High Speed Rail Association
(HSRA)
206 Valley Ct., Suite 800
Pittsburgh, PA 15237
(412) 366-6887

International Mass Transit
Association (IMTA)
P.O. Box 40247
Washington, DC 20016-0247
(202) 362-7960

National Industrial
Transportation League
(NITL)
1090 Vermont Ave. NW,
Suite 410
Washington, DC 20005
(202) 842-3870

The Maintenance Council of the
American Trucking
Accociations (TMC)
2200 Mill Rd.
Alexandria, VA 22314
(703) 838-1763

Transportation Safety
Equipment Institute (TSEI)
300 Sylvan Ave.
P.O. Box 1638
Englewood Cliffs, NJ
07632-0638
(201) 569-8500

Transportation Institute (TI)
5201 Auth Way
Camp Springs, MD 20746
(301) 423-3335

American Public Works
Association (APWA)
1313 E. 60th St.
Chicago, IL 60637
(312) 667-2200

Transportation Research Board
(TRB)
2101 Constitution Ave. NW
Washington, DC 20418
(202) 334-2934

Transportation Research Forum
(TRF)
1600 Wilson Blvd., Suite 905
Arlington, VA 22209
(703) 525-1191

*Transportation Research
Information Service (TRIS)*
Transportation Research
Board
National Research Council
2101 Constitution Ave. NW
Washington, DC 20418
(202) 334-3250

*Journal of Advanced
Transportation*
Institute for Transportation,
Inc.
Duke Station
P.O. Box 4670
Durham, NC 27706
(919) 684-8834

Transportation Journal
P.O. Box 33095
Louisville, KY 40232
(502) 451-8150

Urban Transport News
Business Publishers, Inc.
951 Pershing Dr.
Silver Spring, MD 20910
(301) 587-6300

Census of Transportation
Government Printing Office
Washington, DC 20402
(202) 783-3238

National Transportation Statistics
Government Printing Office
Washington, DC 20402
(202) 783-3238

Toys, Sporting Goods

American Toy Export
Association (ATEA)
c/o Kraemer Mercantile
Corp.
200 5th Ave., Room 1303
New York, NY 10010
(212) 255-1772

Toy and Hobby Wholesalers
Association of America
(TWA)
P.O. Box 955
Marlton, NJ 08053
(609) 985-2878

International Committee of Toy
Industries
c/o David Hawtin
British Toy and
Manufacturers Association
80 Camberwell Rd.
London, England SE5 0EG

Toy Manufacturers of America
(TMA)
200 5th Ave., room 740
New York, NY 10010
(212) 675-1141

National Association of Doll
and Stuffed Toy
Manufacturers, Inc.
(NADSTM)
200 E. Post Rd.
White Plains, NY 10601
(914) 682-8900

Athletic Goods Team
Distributors (AGTD)
1699 Wall St.
Mt. Prospect, IL 60056
(708) 439-4000

Sporting Goods Manufacturers
Association (SGMA)
200 Castlewood Dr.
North Palm Beach, FL 33408
(407) 842-4100

National Association of
Sporting Goods
Wholesalers (NASGW)
P.O. Box 11344
Chicago, IL 60611
(312) 565-0233

International Union of Allied
Novelty and Production
Workers (IUANPW)
181 S. Franklin Ave.
Valley Stream, NY 11581
(212) 889-1212

Antique Toy Collectors of
America (ATCA)
c/o Robert R. Grew
Carter, Ledyard and Milburn
2 Wall St., 15th floor
New York, NY 10005
(212) 238-8803

USA Toy Library Association
(USA-TLA)
2719 Broadway Ave.
Evanston, IL 60201
(312) 864-8240

Sporting Goods Business
Gralla Publications
1515 Broadway
New York, NY 10036
(212) 869-1300

National Sporting Goods
Association
Lake Center Plaza Building
1699 Wall St.
Mt. Prospect, IL 60056
(312) 439-4000

Official Toy Trade Directory
Edgell Communications, Inc.
7500 Old Oak Blvd.
Cleveland, OH 44130
(800) 421-4618 or
(216) 243-8100

Playthings Directory
Geyer-McAllister
Publications, Inc.
51 Madison Ave.
New York, NY 10010
(212) 689-4411

Toy Wholesalers Association of
America
66 E. Main St.
Moorestown, NJ 08507
(609) 234-9155

Leather

Luggage and Leather Goods
Manufacturers of America
(LLGMA)
350 5th Ave., Suite 2624
New York, NY 10118
(212) 695-2340

Luggage and Leather Goods
Salesmen's Association of
America (LLG)
c/o Sara-Jenn Grodensky
370 W. Broadway
Long Beach, CA 11561
(516) 431-4931

Leather Industries of America
(LIA)
1000 Thomas Jefferson St.
NW, Suite 515
Washington, DC 20007
(202) 342-8086

U.S. Hide, Skin and Leather
Association (USHSLA)
1700 N. Moore St., Suite 1600
Arlington, VA 22209
(203) 841-5485

Leathercraft Guild (LG)
P.O. Box 734
Artesia, CA 90701
(213) 864-2420

Leather Workers International
Union (LWIU)
11 Peabody Sq.
P.O. Box 32
Peabody, MA 01960
(508) 531-6200

Leather Industry Statistics
Leather Industries of America
1000 Thomas Jefferson St. NW,
Suite 515
Washington, DC 20007
(202) 342-8086

Jewelry, Silverware

Jewelers of America
Time-Life Building
1271 Avenue of the Americas
New York, NY 10020
(212) 489-0023

Jewelers Board of Trade (JBT)
P.O. Box 6928
Providence, RI 02940
(401) 438-0750

Fashion Jewelry Association of
America (FJAA)
Box S-8, Regency East
One Jackson Walkway
Providence, RI 02903
(401) 273-1515

Jewelers Shipping Association
(JSA)
125 Carlsbad St.
Cranston, RI 02920
(401) 943-6020

Manufacturing Jewelers and
Silversmiths of America
(MJSA)
100 India St.
Providence, RI 02903-4313
(401) 274-3840

American Diamond Industry
Association (ADIA)
71 West 47th St., Suite 705
New York, NY 10036
(212) 575-0525

American Gem and Mineral
Suppliers Association
(AGMSA)
P.O. Box 741
Patton, CA 92369
(714) 885-3918

American Gem Society (AGS)
5901 W. 3d St.
Los Angeles, CA 90036
(213) 936-4367

American Gem Trade
Association (AGTA)
181 World Trade Center
P.O. Box 581043
Dallas, TX 75258
(214) 742-GEMS

Jewelry Manufacturers
Association (JMA)
475 5th Ave., Suite 1908
New York, NY 10017
(212) 725-5599

Jewelry Manufacturers Guild
(JMG)
P.O. Box 46099
Los Angeles, CA 90046
(714) 769-1820

Gemological Institute of
America (GIA)
1660 Stewart St.
Santa Monica, CA 90404
(213) 829-2991

Independent Jewelers
Organization (IJO)
2 Railroad Pl.
Westport, CT 06880
(203) 226-6941

Jewelry Industry Council (JIC)
8 W. 19th St., 4th floor
New York, NY 10011
(212) 727-0130

Jewelry Industry Distributors
Association (JIDA)
120 Light St.
Baltimore, MD 21230
(301) 752-3318

Society of American
Silversmiths (SAS)
P.O. Box 3599
Cranston, RI 02910
(401) 461-3156

Giftware News
Talcott Communications
Corp.
1414 Merchandise Mart
Chicago, IL 60654
(312) 670-0800

China, Glass and Giftware
Association
1115 Clifton Ave.
Clifton, NJ 07013
(201) 779-1600

American Jewelry Manufacturer
Chilton Book Co.
Chilton Way
Radnor, PA 19089
(800) 345-1214 or
(215) 964-4000

Modern Jeweler, National Edition
Vance Publishing Corp.
7950 College Blvd.
Shawnee Mission, KS 66201
(913) 451-2000

Current Business Reports:
Monthly Retail Trade
U.S. Bureau of the Census
Government Printing Office
Washington, DC 20402
(202) 783-3238

American Jewelry Marketing
    Association
1900 Arch St.
Philadelphia, PA 19103
(215) 564-3484

Confidential Reference Book of the
    Jewelers Board of Trade
Jewelers Board of Trade
70 Catamore Blvd.
East Providence, RI 02914
(401) 438-0750

Index

About the Author

Robert J. Boxwell, Jr. is a founding partner of Churchill & Company, a San Francisco–based management consulting firm specializing in benchmarking and strategic and quality advisory services. He has completed a wide variety of benchmarking studies for many of America's best-known companies and has educated hundreds of managers throughout the world on the benchmarking process. He also co-developed and currently teaches the benchmarking course for the American Society for Quality Control; for the University of California, Berkeley; and in-house for a number of *Fortune* 500 and international companies.